D0116761

CENOZOIC HISTORY AND PALEOCEANOGRAPHY OF THE CENTRAL EQUATORIAL PACIFIC OCEAN

The Geological Society of America
Memoir 143

Cenozoic History and Paleoceanography of the Central Equatorial Pacific Ocean

A Regional Synthesis of Deep Sea Drilling Project Data

TJEERD H. VAN ANDEL
G. ROSS HEATH
T. C. MOORE, JR.

School of Oceanography
Oregon State University
Corvallis, Oregon 97331

1975

To the memory of Fritz F. Koczy,
first chairman of the Planning Committee of JOIDES,
who more than most of us helped to change
deep-ocean drilling from a dream to reality

The Memoir series was originally made possible
through the bequest of
Richard Alexander Fullerton Penrose, Jr.

Published by
THE GEOLOGICAL SOCIETY OF AMERICA, INC.
3300 Penrose Place
Boulder, Colorado 80301

Printed in the United States of America

Contents

Figure

Preface and Acknowledgments

Since 1968, cruises of the D.V. *Glomar Challenger* have created a vast and immensely valuable collection of core samples and geologic data for hundreds of drill sites in all oceans and in the Caribbean, Mediterranean, and Red Seas. This collection represents a large part of the Cenozoic and Mesozoic record of the oceans. The material enables us to reconstruct the geologic history of the ocean basins and improve our understanding of the history of the Earth, the evolution of life, temporal changes in oceanic circulation, and indirectly, the evaluation of the Earth's climate. Regional syntheses are indispensable for effective planning for future ocean drilling. Our study provides a synthesis of the data available for the central equatorial Pacific Ocean, with emphasis on the depositional history of pelagic biogenic sediment and its control by tectonic events and paleoceanographic conditions.

Because of our intimate involvement in the Deep Sea Drilling Project from the earliest stages, the completion of this study gives us particular satisfaction. We could not have achieved this without the assistance of many people. The study is based mainly on published data of the Deep Sea Drilling Project and, hence, directly on the efforts of shipboard scientists, technicians, drillers and crew, the personnel of the shore laboratories, and the staff of the Deep Sea Drilling Project. With the help of David Bukry, William R. Riedel, Anika Sanfilippo, and Jean Westberg, we have been able to use the latest revisions of the biostratigraphy. The staff at the Deep Sea Drilling Project, in particular, Lilian B. Musich, William R. Riedel, Peter R. Supko, Peter B. Woodbury, and Paula Worstell, has given us prompt and often extensive assistance in furnishing samples, extracting additional information from prime data files, and tracking down elusive suspected errors.

At Oregon State University, J. Paul Dauphin, Kenneth Keeling, Natasa Sotiropoulos, Kathryn Torvik, Linda Wille, and Jody Imlah have contributed in the laboratory analysis, data processing, drafting, and manuscript preparation. We have had the benefit of extensive discussions and constructive criticism from many colleagues, in particular, Margaret Leinen, Nicklas G. Pisias, and Bruce T. Malfait. Wolfgang H. Berger, William A. Berggren, James D. Hays, Bilal ul Haq, David A. Johnson, and James P. Kennett undertook the arduous task of reviewing the bulky manuscript and improved it greatly in doing so. To all of them and to many others whom we have not named we express our warmest thanks.

The investigation was supported by National Science Foundation Grant GA-31478.

Abstract

This study has three distinct but interrelated objectives: to prepare a geological synthesis of Deep Sea Drilling Project data from the central equatorial Pacific Ocean, to interpret this information in terms of the paleoceanographic history of this region, and to evaluate the usefulness of drill data and develop procedures and strategies for future studies of this kind. The investigation is based on primary data contained in the *Initial Reports of the Deep Sea Drilling Project* and is supported by information from surface cores. The principal data used are the biostratigraphy, lithology, carbonate content, bulk density, and porosity of the cores. From these properties, sedimentation rates, carbonate and carbonate-free accumulation rates, and paleobathymetric histories of the drill sites were derived with the aid of Berggren's chronology. Paleopositions of the drill sites and surface cores were determined from rotation parameters of the Pacific and Cocos plates.

The present surface and deep circulation, fertility patterns, and sedimentation of the equatorial Pacific constitute a frame of reference for the paleoceanographic evolution. East-west and north-south lithologic profiles show that a zone of maximum deposition approximately parallel to the Equator has existed at least since middle Eocene time. With increasing age, the axis of this zone is found progressively farther north of the Equator. The profiles illustrate a gradual change from calcareous to siliceous deposits with increasing depth at any time, and they indicate an abrupt change from a dominantly siliceous to a dominantly calcareous depositional regime at the Eocene-Oligocene boundary. Large changes with time in the width and sedimentation rate of the calcareous equatorial zone indicate major variations in depositional conditions since Eocene time.

Subsidence with age of the oceanic basement, plate rotation, and changes in spreading rate are closely examined in this study. A northward shift of the equatorial zone of maximum deposition with age and trend and ages of linear volcanic island chains (melting spots) define the rotation of the Pacific plate. This gives a pole of rotation at lat 67° N, 59° W, and rotation rates of 0.25°/m.y. before and 0.83°/m.y. after 25 m.y. B.P. This rotation scheme describes the migration with time of the drill sites and, in combination with the subsidence histories of the drill sites, permits the reconstruction of the paleobathymetric evolution of the region. About 50 m.y. ago, the eastern edge of the Pacific plate was located at long. 115° W and migrated rapidly eastward during the next 20 to 30 m.y., becoming stationary at approximately long. 100° to 105° W, notwithstanding a large crestal jump from the Mathematicians-Clipperton ridge system to the present East Pacific Rise 10 to 15 m.y. ago. The ancestral East Pacific Rise was a relatively narrow, symmetric feature with steep upper slopes. About 25 m.y. ago, it developed a much broader and gentler west flank, thereby acquiring its present asymmetry. The early rise crest was much shallower than today, perhaps suggesting an increase in spreading rate about 35 to 40 m.y. ago. At present, a broad shoal region near 4,000 m extends far west of the rise crest south of lat 10° S. Paleobathymetric and sedimentologic evidence indicates that during middle Tertiary time this shoal was located much farther north.

The history of the calcite compensation depth (CCD) is developed on the basis of the mean carbonate content for each 1-m.y. interval in all drill sites and the carbonate content of surface cores that penetrate pre-Quaternary sediment. The data show that the history of the CCD in a narrow equatorial zone was different from the history of the Pacific CCD more than 4° of latitude away from the Equator. During the Eocene the CCDs in both the equatorial and Pacific regions were shallow and located near 3,000 m, but about 38 m.y. ago they dropped precipitously, the equatorial CCD to about 5,000 m and the Pacific CCD to 4,300 m. During the latter part of the Oligocene the Pacific CCD began to rise again, reaching a minimum at about 3,800 m in late Miocene time. The equatorial CCD began to rise much later in late Miocene time, reaching a minimum between 4,700 and 4,400 m about 6 to 7 m.y. ago before dropping to the present level.

The depositional history in the equatorial region is developed with the aid of isopleth maps of carbonate and carbonate-free accumulation rates. The results are sensitive to errors in the absolute chronology, but with the exception of a minor modification for the period 10 to 16 m.y. B.P., the adopted chronology produces consistent results. The evidence points to a series of major changes in sedimentation-rate patterns with time. During Eocene time, the equatorial zone was narrow, of restricted westward extent, and characterized by low carbonate accumulation rates. At 38 m.y. B.P., a change began that led to a very broad equatorial zone with high carbonate accumulation rates that extended westward well beyond even its present western limit. Since middle Miocene time, the width and accumulation rate of the equatorial zone decreased again. The changes parallel depth changes in the equatorial CCD but are not always synchronous. The carbonate-free accumulation rate increased gradually with time with a few pronounced deviations, especially during late Miocene time. A marked correlation between paleodepth and carbonate accumulation can be demonstrated but cannot, by itself, explain the major variations with time. Superimposed on this general pattern are shorter term fluctuations in sediment composition and sedimentation rates for 3- to 10-m.y. intervals and very short term fluctuations on the order of a few tens to a few hundreds of thousands of years. These shorter term fluctuations may be climatically controlled, but compositional and chronologic data are inadequate to examine their origin properly.

Several major erosional phases occurred in the deep Pacific during the past 50 m.y. The earliest one reached its maximum extent about 42 m.y. ago, and the distribution of hiatuses suggests that it resulted mainly from bottom water flowing into the Pacific from the tropical Indian Ocean. Another erosional period around 30 to 35 m.y. ago was restricted to the western Pacific and did not interrupt the depositional regime in the equatorial zone. A major erosional event in the equatorial Pacific reached a maximum about 12 m.y. ago; it may be attributed to enhanced bottom-water flow that resulted from major development of antarctic glaciation. The latest equatorial phase, coinciding with the beginning of arctic glaciation 2 to 3 m.y. ago, did not greatly affect the southern and western Pacific. Thus, of the major erosional phases in the equatorial Pacific, the Eocene phase was associated with a change in bottom-water flow, whereas the 12-m.y. phase correlates with the formation of the antarctic ice cap.

Variations in carbonate accumulation rate with paleodepth are used to establish the change with time of the vertical gradient of the carbonate dissolution rate. This gradient appears to have responded mainly to changes in the structure of the deep and bottom waters, whereas the CCD was primarily determined by the global carbonate budget and by changes in the locus of carbonate deposition through time. After changes in carbonate dissolution are taken into account, variations in the carbonate and carbonate-free accumulation rates allow assessment of changes in upwelling and fertility.

The paleoceanographic indicators resulting from the study, in context with data from other regions, yield the following oceanographic history for the past 50 m.y. During the initial phase, prior to 38 m.y. B.P., carbonate supply was low, and dissolution was extensive. This resulted in a narrow carbonate zone and low rates of accumulation. Silica was mobilized to form chert, and erosion became widespread. Around 38 m.y. B.P., carbonate input increased abruptly, and solution decreased markedly. As a result, the equatorial carbonate zone widened greatly. Bottom-water characteristics may have changed owing to the first extensive development of sea ice around Antarctica. About 33 m.y. B.P., dissolution rates at depth began to increase again, but this increase was compensated for either by a large increase in the carbonate supply or by a depression of the lysocline. As a result, until about 26 m.y. B.P., a very broad and extensive equatorial carbonate zone with maximal accumulation rates existed. The steepening of the dissolution gradient may be related to a decrease in influx of bottom water resulting from major hydrographic changes around Antarctica. The increase in carbonate supply may have resulted from final closure of the Tethys seaway and the ensuing narrowing and intensification of equatorial upwelling. About 15 m.y. B.P., the rate of carbonate dissolution increased further, and the CCD shoaled. As a result, the equatorial carbonate zone became much narrower, and many hiatuses were formed. At this time, the lysocline may have attained its present position, which suggests that the present configuration of sources and mechanisms of intermediate- and deep-water supply to the Pacific was being developed. This probably resulted from the major development of antarctic glaciation that began during this period and increased the supply of bottom water. Some evidence from the carbonate and carbonate-free accumulation rates also suggests an increase in the fertility of the equatorial region, perhaps resulting from closure of the western Pacific Ocean and the formation of the Cromwell Current at this time. About 3 to 4 m.y. B.P., the onset of arctic glaciation marked the beginning of large and rapid changes in depositional conditions that cannot be followed in detail with the Deep Sea Drilling Project data.

1

Scope and Objectives

Sedimentation rates in the pelagic realm are a function of a complex set of variables. Subsidence of the oceanic crust with increasing age, changes in the crestal depth of mid-ocean ridges with time, plate migration across latitudinal and longitudinal zones, and temporal variations in biologic productivity, supply of nonbiogenic sediment, calcium carbonate dissolution with depth, and deep-sea erosion are just a few of the major influences on rates of deposition of pelagic sediment. From the beginning, results of the Deep Sea Drilling Project (Maxwell and others, 1970) have shown that these effects are often of such magnitude that long-term sedimentation rates cannot be reasonably estimated without additional evidence.

Several basic causes underlie temporal and spatial changes of sedimentation rates: some are the result of plate motions, others are caused by changes in the surface and deep circulation of the oceans. Still others can be attributed to changes in the response of planktonic life (the main source of pelagic sediment) or to environmental or evolutionary changes. In addition, the influx of nonbiogenic sediment, such as terrigenous clay and volcanic ash, may vary as a result of tectonic, orographic, and hydrologic changes with time.

The possibility of a complex history of pelagic sedimentation and the need to examine it in detail were recognized in the initial proposal for the Deep Sea Drilling Project, which stated as the second principal objective the need to attain "a paleo-oceanographic insight into the history of the major current systems and water masses which can be deduced from the patterns of distribution of microfossils and other sediment characteristics on various time planes, and of its consequences for the history of climate" (proposal by Scripps Institution of Oceanography to the National Science Foundation, dated August 23, 1965). Initially, a meridional traverse located at long. 140° W in the Pacific Ocean was drilled for this purpose. Later, numerous other sites were cored in the same central equatorial region. At the present time, approximately 25 drill holes from Legs 5, 8, 9, 16, and 17 of the cruise of the D.V. *Glomar Challenger* make this region one of the most intensively drilled parts of the ocean basins. Furthermore, numerous cruises by oceanographic vessels of the Scripps Institution of Oceanography and Lamont-Doherty Geological Observatory have provided a large collection of surface cores, many of which penetrated into pre-Quaternary deposits. Because the plate tectonic history of the central equatorial Pacific is relatively simple, this region provides an excellent target for a synthesis of the sedimentation history of a large oceanic area. The global significance of such a synthesis is enhanced by the fact that the region straddles one of the Earth's most highly productive oceanic regions, the equatorial current system, and therefore, the results have an important bearing on the paleocirculation of the oceans.

In the design of this study, four principal control factors for the processes and products of sedimentation were stressed: northward migration of the Pacific plate across the Equator, subsidence of the oceanic basement with increasing age away from the East Pacific Rise, changes with time in the depth profile of calcium carbonate solution, and changes with time in the amount of biogenic sediment produced in the equatorial current system. Initially, we thought that a dynamic model of sedimentation could be devised that would reflect these controls and could be tested against observed sediment-thickness variations (Ewing and others, 1968), thus furnishing an independent test of the validity of the inferred temporal variations of the basic factors. It soon became apparent that the paleoceanographic history of the central equatorial Pacific was much more complicated than foreseen and that such a model could not be usefully constructed or used to test our inferences, and this approach was abandoned.

The scope and focus of our study are mainly determined by the data available in the *Initial Reports of the Deep Sea Drilling Project*, although we have also drawn extensively on the literature and to some extent on unpublished data from surface cores in the region. The information readily available in the *Initial Reports* consists of biostratigraphic age data, lithologic descriptions, quantitative measurements of the calcium carbonate content, and rates of deposition based on age and sediment thickness. Using measurements of wet-bulk density and porosity, also included in the *Initial Reports*, the latter can be converted into quantitative rates of accumulation expressed as the weight of sediment accumulating on a unit of area in a unit of time.

Two principal types of sedimentary material, consisting respectively of the calcite skeletons of foraminifers and nannoplankton and of the opaline tests of radiolarians and diatoms, are produced in the surface waters of the oceans. These two components react quite differently during deposition. The calcium carbonate is strongly dissolved below the lysocline and is entirely removed where the sea-floor depth exceeds the calcite compensation depth, thus showing a strong negative correlation with the depth of deposition. Opal, although also strongly dissolved in the water column and at the sea floor, shows no correlation with depth, and an opal compensation depth does not exist (Heath, 1974). As a result, separate rates of deposition for calcium carbonate and opal should provide a means of distinguishing the effects of changes in biological productivity and in carbonate dissolution with time. Unfortunately, opal determinations are not part of the *Initial Reports*, and the carbonate-free residue values that can be derived from the data include terrigenous clay and quartz, volcanic ash, and other nonbiogenic components that may mask variations in opal content. In a separate study based on new determinations of opal and terrigenous mineral content (M. Leinen, in prep.), this subject is pursued further than it can be here.

Geographically, our study is limited by several considerations. The main concentration of drill sites lies between lat 15° N and 15° S and long. 150° W and 95° W. North of lat 15° N, drill sites have mainly penetrated brown clay that yields insufficient biostratigraphic age information; south of lat 15° S no long sections have been cored. East of long. 95° W several good sections are available, but the tectonic history of this region, which comprises the zone of interaction between the Pacific, Cocos, and Nazca plates, is complex (Rea and Malfait, 1974) and is too controversial to permit a reasonable reconstruction of plate migrations. In the west, near the Line Islands, there are several drill sites, but the post-Cretaceous sections are short and incomplete, and the determination of basement ages is uncertain owing to the low magnetic latitude and the possibility of late volcanic intrusions related to the Line Islands volcanic chain (Winterer, 1973).

We have restricted the investigation to the past 50 m.y. B.P. Older sections are found only at the western edge of the area; core recovery of the earliest Cenozoic deposits is generally poor because of the occurrence of numerous chert beds; and the precision of absolute-age assignments deteriorates markedly for sediments older than about 45 to 50 m.y.

The procedures used in processing the data and the processed data are presented in detail in the appendixes. In the following chapters, we discuss the present oceanographic conditions and the modern depositional regime of the central equatorial Pacific (Chap. 2). In Chapter 3, the distribution of sediments in space and time will be presented, followed by a review of the tectonic history of the region and its consequences for the migration of the drill sites across the Equator and for the physiographic history of the area. The history of the dissolution of calcium carbonate with depth, based on data from drill sites and surface cores that penetrated pre-Quaternary deposits, is discussed in Chapter 4. Chapter 5 deals with rates of deposition and their interpretation in terms of temporal and latitudinal changes in carbonate solution and in production of biogenic material. In Chapter 6, the evidence for phases of deep-sea erosion is examined, and in Chapter 7 we attempt to interpret all data in a paleoceanographic context.

Much available information regarding the Cenozoic history of the central equatorial Pacific Ocean has not been used in this synthesis. Except for biostratigraphic purposes, we have made minimal use of the paleoecologic information that is potentially available in numerous studies on fossil assemblages contained in the *Initial Reports* and elsewhere. The *Initial Reports* also contain extensive mineralogic data obtained by x-ray diffraction. Unfortunately, the mineralogic data are of little value for the central equatorial Pacific (and for many other oceanic regions), because the high opal content of the carbonate-free fractions masks most of the minerals that are useful provenance indicators. Moreover, the lack of an internal standard in the analyses prevents the conversion of relative mineral percentages to absolute concentrations for use in the form of rates of accumulation. Attempts to extract meaningful patterns from such data in the *Initial Reports* have thus far been futile.

In concentrating on the carbonate content of the sediment and on the rates of deposition, we have also neglected other sedimentologic data in the *Initial Reports* or obtainable by analysis of core samples. We have done so in part because we wished to limit the scope of this already extensive study and in part because use of this type of information would have required much additional laboratory work that was beyond the time limits set for this study. We hope that the stratigraphic and paleoceanographic framework furnished by our study will facilitate such investigations.

2

Present Oceanic Circulation
and Sedimentation

The equatorial Pacific Ocean is one of the Earth's most biologically productive ocean regions (Fig. 1; Reid, 1962; Ryther, 1963; Lisitsin, 1970; Koblentz-Mishke and others, 1970). Upwelling associated with the equatorial current system (Fig. 2) provides nutrient-rich water that supports a primary productivity of 200 to 500 mg $C/m^2/day$ (Koblentz-Mishke and others, 1970), which in turn produces a high zooplankton standing crop (Fig. 3).

Siliceous and calcareous tests secreted in this region of high biological productivity dominate the equatorial sedimentation pattern. The presence of a broad zone of highly fossiliferous sediment along the Equator has been known for almost a century (Murray and Renard, 1891; Revelle, 1944; Arrhenius, 1952; Bramlette, 1961). In deep water west of the crest of the East Pacific Rise, these deposits grade from highly calcareous near the Equator (Fig. 4) to dominantly siliceous to the north and south. At even greater distance, the siliceous deposits grade into barren brown

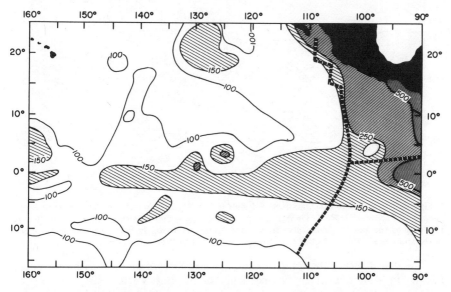

Figure 1. Primary productivity (g $C/m^2/yr$) in the eastern equatorial Pacific, after Koblentz-Mishke and others (1970) and Moore and others (1973). Heavy dashed line is crest of East Pacific Rise and Galapagos rift zone.

9

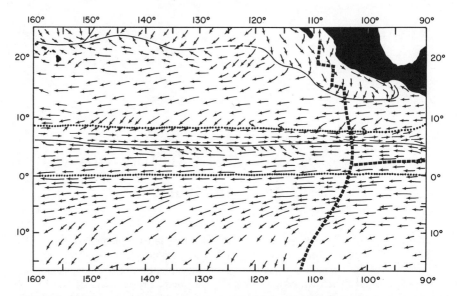

Figure 2. Surface currents in the eastern equatorial Pacific (Northern Hemisphere winter) after Defant (1961). Dotted lines show regions of divergence (upwelling). Heavy dashed line is crest of East Pacific Rise and Galapagos rift zone. Other dashed and straight lines are boundaries of current systems.

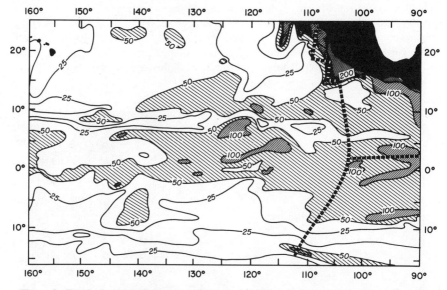

Figure 3. Zooplankton displacement volumes (ml/1,000 m³) in surface water of the eastern equatorial Pacific, after Reid (1962). Although indicative of standing stock rather than production, this map is the best available clue to the distribution of living foraminifers and radiolarians in the study area. Heavy dashed line is crest of East Pacific Rise and Galapagos rift zone.

("red") clay beneath the infertile central water masses of the North and South Pacific. In shallower water, on the flanks of the East Pacific Rise close to the crest, the calcareous zone extends farther north and south than in the deeper water to the west.

The areal distribution of calcium carbonate in surface sediments reflects the

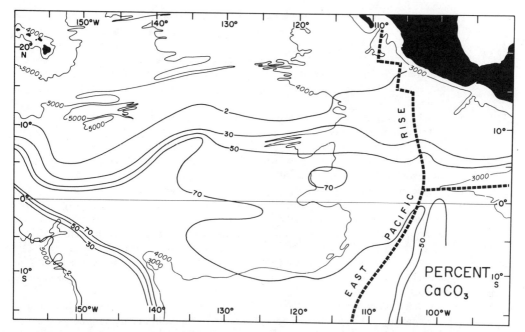

Figure 4. Calcium carbonate distribution in Holocene sediment of the eastern equatorial Pacific, modified from Lisitsin (1970) and Bramlette (1961) with additional data. Contours are percentages by weight. Heavy dashed line is crest of East Pacific Rise and Galapagos rift zone.

interplay between the rate of biological fixation of calcareous material in surface water and the rate of solution of calcium carbonate at depth (Arrhenius, 1963). Below the upper few hundred metres, Pacific Ocean waters are undersaturated with calcium carbonate, so that calcareous skeletal material tends to dissolve as it settles through the water column, especially during exposure near the sediment-water interface. In the central equatorial Pacific the rate of solution increases rapidly at a depth of about 3,700 m (Peterson, 1966; Berger, 1970a), with a concomitant large change in the foraminiferal and coccolith composition of the calcareous material (Berger, 1973b). This zone of rapid faunal change was named the "lysocline" by Berger (1968). In the equatorial Pacific, west of the East Pacific Rise, the lysocline is found at a depth of 3,700 to 4,000 m. Below this depth, the rate of solution of calcium carbonate appears to increase linearly with depth (Heath and Culberson, 1970) until the rates of solution and supply are equal at the calcite compensation depth (CCD). Below this level, sediment is free of carbonate. Berger and Winterer (1974) showed that in the central equatorial Pacific between approximately lat 7° N and 4° S, the CCD is somewhat below 5,000 m. To the north and south it rises, reaching 4,500 m at approximately lat 15° N and 10° S. The equatorial depression of the CCD results from a large excess influx of calcareous sediment (Arrhenius, 1963).

The distribution of siliceous tests, reflected in the opal content of carbonate-free sediment (Fig. 5), is essentially independent of depth. Although opal dissolves in the water column and in surface sediment, the degree of undersaturation and the rate of dissolution are fairly uniform in deep and bottom waters. Consequently, opal abundance patterns mirror surface productivity much better than does the carbonate pattern. The Equator is marked by a narrow belt of highly siliceous sediment under the equatorial divergence. This belt trends slightly north of west, probably reflecting the intense westerly flowing Peru Current, which joins the equatorial current system farther east than does the corresponding California Current to the north.

The decreasing concentration gradient east of about long. 110° W largely reflects dilution by hemipelagic sediments from Central and South America. The decreasing gradient to the west corresponds to a general east-to-west decrease in surface biological productivity (Fig. 1) and probably results from a decreased rate of opal supply superimposed on a relatively uniform influx of nonbiogenous sediments.

Thus, in deep water, the northward and southward transitions from calcareous to siliceous deposits result from the superimposition of a depth-dependent solution process onto a productivity gradient. The transition from siliceous ooze to brown clay is mainly due to the decline in productivity away from the Equator.

Calcareous material constitutes a large part of the total sediment supply (Lisitsin, 1972). Consequently, its partial or complete removal by solution sharply reduces the rate of sediment accumulation on the ocean floor. In the central equatorial Pacific, recent calcareous sediments in the equatorial zone accumulate at rates of 10 to 20 mm/1,000 yr depending on water depth, whereas biogenic siliceous ooze accumulates at about 4 to 5 mm/1,000 yr. In contrast, the almost fossil-free sediments at higher latitudes in the North and South Pacific accumulate at rates of 2 mm/1,000 yr or less.

If these conditions have been stable for tens of millions of years, they should have produced a body of sediment that is lens-shaped in north-south cross section, with the thickest part lying on or slightly north of the Equator (Arrhenius, 1952). The thickness of sediment determined from seismic profiles appears to confirm the existence of such a lens (Ewing and others, 1968). The section decreases in thickness from more than 0.5 sec (two-way travel time) near the Equator to less than 0.1 sec near lat 10° N and 10° S. However, in many areas west of the East Pacific Rise, thickness measurements from seismic profiles are limited by an opaque reflector above basement that masks the underlying sediment. Deep-sea drilling

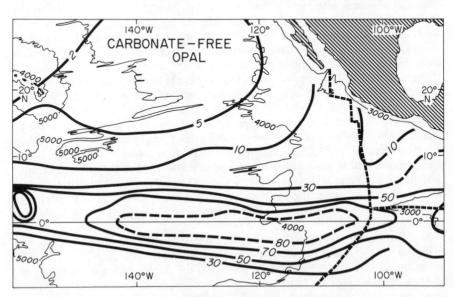

Figure 5. Opal distribution in Holocene sediment of the eastern equatorial Pacific, after Heath and others (1975). Contours are percentages by weight of carbonate-free sediment. Heavy dashed line is crest of East Pacific Rise and Galapagos rift zone.

has shown that this opaque layer consists of numerous chert beds of Eocene and Paleocene age (Tracey and others, 1971). As a result, the isopach map of Ewing and others (1968) shows only minimum sediment thicknesses. With increasing distance from the crest of the East Pacific Rise and increasing age of the oceanic crust, the isopach map becomes progressively more distorted and reflects only post-Eocene sedimentation. In Figure 6, the isopach map based on seismic profiling is compared with one corrected by means of DSDP data. The existence of a lens-shaped deposit, with its long axis parallel to and slightly north of the Equator, is clearly confirmed, but the drill data reveal a large bulge extending northwestward toward Hawaii which is not seen on the seismic profiles and consists largely of pre-Oligocene deposits. The bulge suggests that depositional conditions are complex and not simply controlled by equatorial productivity and calcium carbonate dissolution.

In general, the isopach map and the map of calcium carbonate abundance in surface sediments are similar, owing to the high rate of accumulation of calcareous sediments. On the crest of the East Pacific Rise, however, the relationship does not hold because of the very young crustal age. Furthermore, south of lat 10° S and between long. 120° W and 140° W, highly calcareous sediments occur at shallow depth on old crust far west of the East Pacific Rise. The thickness of sediment here is quite small, however.

Outcrops of pre-Quaternary sediment in the central Pacific both north and south of the Equator have been known for more than 20 yr (Arrhenius, 1952). Surface deposits in a narrow zone on the Equator are generally of Quaternary age, but sediment of Tertiary age is commonly sampled a few degrees farther north and, to a lesser extent, south. Increasingly older deposits crop out with increasing distance from the Equator (Riedel and Funnell, 1964; Hays and others, 1969; Burckle and others, 1967). Erosion and slumping on young fault scarps can be invoked to explain the local exposure of older sediments (Moore, 1970; Johnson, 1972b), because little sediment needs to be removed in order to expose deposits as old as late Oligocene. However, the exposure of Eocene and lower Oligocene sediments would have required the erosion of tens to a hundred metres of overburden, and the areal extent of the outcrops indicates large-scale erosion. The recognition of laterally transported abyssal sediment and evidence for topographically controlled deposition indicate erosional processes of regional extent (Johnson and Johnson, 1970; Johnson, 1972a, 1972b). This erosion may be produced by Antarctic Bottom Water flowing northeastward into the central and northern Pacific Ocean through deep channels between ridges and seamount chains. Details of the intensity and distribution of this regional erosion and its time of initiation are not well understood. Erosion, or at least nondeposition, may have commenced even before earliest Miocene time or the entire process may have occurred during the Quaternary.

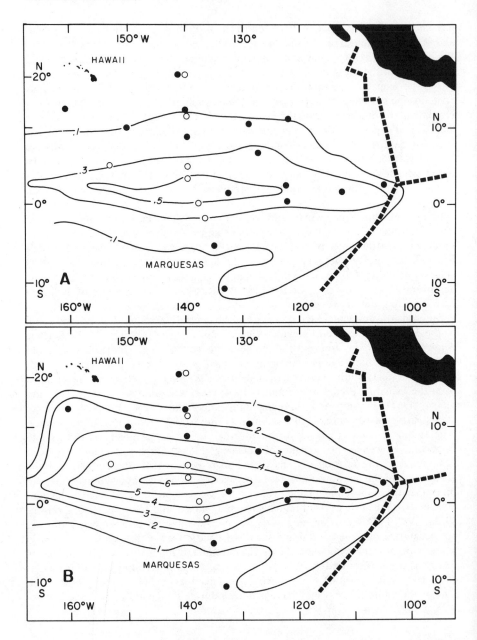

Figure 6. Isopach maps of sediment thickness in the central equatorial Pacific. A, isopachs in seconds of two-way travel time (approximately 1,000 m/sec), after Ewing and others (1968). B, isopachs in metres; modified from A by correction with data from drill sites. Difference between A and B is mainly due to the inability of the profiler system to detect basalt basement below a thick zone of early Cenozoic chert. Black dots are drill sites where basement was reached; open circles are drill sites terminating in sediment. Heavy dashed line is crest of East Pacific Rise and Galapagos rift zone.

3

Regional Tectonic History and Geology

TECTONIC EVOLUTION

Four lithospheric plates compose the central equatorial Pacific Ocean: the Pacific plate in the west, and the North American, Cocos, and Nazca plates in the east (Fig. 7). The plate boundary of greatest importance for this study lies along the East Pacific Rise, where new crust is being formed at a rate of 8 to 20 cm/yr. This crust is subducted in the western Pacific and along the margin of Central and South America. The Cocos and Nazca plates, separated by an east-trending accreting plate boundary, are remnants of a former, much larger Farallon plate. This plate was consumed off the North American coast as far south as the Gulf of California between 30 and 10 m.y. ago (Atwater, 1970). The remaining southern part began to divide with the formation of the Galapagos rift zone 30 to 40 m.y. ago (van Andel and others, 1971; Hey and others, 1972; Heath and van Andel, 1973; Rea and Malfait, 1974).

All drill sites used in this study except one are on the Pacific plate. The history of this plate during the past 50 m.y. is of fundamental importance in the unraveling of the equatorial depositional history. The history of the Cocos and Nazca plates and the motions of these plates relative to the Pacific plate, on the other hand, are of less direct importance. Three aspects of the history of the Pacific plate are particularly relevant. During the Cenozoic, the plate continuously changed position relative to the Equator. A rotational scheme must be determined that permits reconstruction of the migration of the drill sites. Secondly, spreading from the East Pacific Rise and its precursors has resulted in a westward increase in age and subsidence of the sea floor. Information on the spreading history is therefore required to permit use of drill sites in which basement was not reached and of surface cores penetrating pre-Quaternary deposits. This information is also needed for the reconstruction of the bathymetric histories of the drill sites. Finally, the location of the East Pacific Rise has not remained fixed throughout the Cenozoic, and because of a large effect on the paleobathymetric reconstruction, its history needs to be determined.

Sclater and others (1971) and Herron (1972) have shown that the East Pacific Rise is a relatively young feature. Formerly, the eastern boundary of the Pacific plate south of lat 4.5° S,was located at the Galapagos Rise (Fig. 7). North of lat 4.5° S, Herron (1972) postulated, on the basis of magnetic anomaly patterns, a fossil ridge between lat 7° N and the Equator. Other fossil ridge crests exist between the Rivera and Orozco Fracture Zones (Mathematicians Ridge) and south of the Clipperton Fracture Zone (Clipperton Ridge) (Sclater and others, 1971). The existence of a fossil crest between the Orozco and Clipperton Fracture Zones is in doubt; its presence is suggested by a north-south zone of rough terrain and

15

seamounts, but Anderson and Davis (1973) have argued that the age-depth relationships do not support this interpretation. Such rather large shifts in position of the plate edge may not be unique; Anderson and Sclater (1972) used age-depth relationships to postulate several crestal jumps on the flanks of the East Pacific Rise south of lat 5° S, thereby implying that this behavior may have been common throughout the history of the eastern Pacific plate boundary (Rea, 1974). If this is true, it might explain the large and abrupt variations in spreading rate on the Pacific plate north of the Equator (Winterer, 1973).

The age of the fossil ridge system has not been established firmly. Herron (1972) assumed that the entire older system became extinct about 9 m.y. ago, but that for some time prior to then both the older and newer systems coexisted, with the East Pacific Rise steadily extending northward. Sclater and others (1971), using depth-age relationships, assumed an extinction at 4 m.y. B.P. for the Mathematicians-Clipperton system, whereas Anderson and Sclater (1972) gave a similar time for the extinction of the Galapagos Rise and an age of 6.5 m.y. for the East Pacific Rise between lat 5° S and 15° S.

The age and location of the ancestral East Pacific Rise have important consequences for the absolute migrations assumed for drill sites 81 and 82 (Fig. 7),

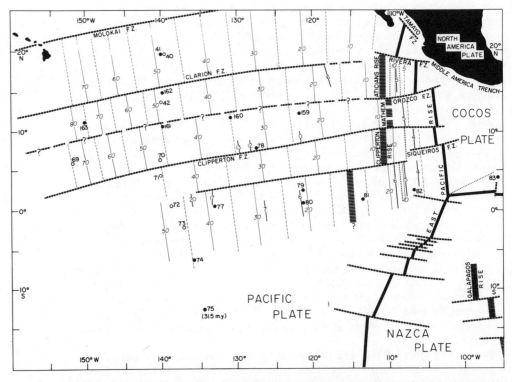

Figure 7. Tectonic and basement isochron chart of the central equatorial Pacific. Tectonics after Sclater and others (1971), Herron (1972), and Anderson and Sclater (1972) with modifications. Heavy line is present ridge crest; broad striped zone is extinct ridge crest; heavy dashed lines are fracture zones. Very thin lines with numbers 10 through 70 are basement isochrons. Heavier lines numbered 4, 5, 6, 7, 17, and 20 are magnetic anomalies after Herron (1972). Approximate boundary of crust generated from present and extinct ridge system is shown by dotted line between crests. Black dots are drill sites where basement was reached; open circles are drill sites terminating in sediment. See Appendix 2 for alternate interpretations of isochrons north of Clarion Fracture Zone. Basement age is shown for site 75.

which are, respectively, 16.5 and 5.6 m.y. old (van Andel and Bukry, 1973). Site 83, with a crustal age of 8.2 m.y., is not affected, because it has been located continuously on the Cocos plate. If the fossil ridge became extinct 10 m.y. ago, only site 81 has participated in the motions of two successive plates, whereas site 82 has always moved with the Pacific plate. The question cannot be resolved unequivocally because of the uncertainties involved in the depth-age argument used by Sclater and others (1971), Anderson and Sclater (1972), and Anderson and Davis (1973) and because of the difficulty of interpreting magnetic anomalies at this low latitude (Herron, 1972). For the present study we have assumed, in accordance with Herron's view, that the fossil ridge between lat 5° N and the Equator became extinct 11 to 12 m.y. ago. This interpretation is supported by Figure 8, although Mammerickx and others (1975) argued in favor of extinction about 6 m.y. ago. Spreading symmetry and a more uniform spreading rate are obtained by assuming that the Mathematicians Ridge ceased to function approximately 5 to 7 m.y. ago, the Clipperton Ridge 8 to 9 m.y. ago, and the ridge between lat 5° N and the Equator 10 to 12 m.y. ago. This progressive northward extinction in time thus parallels the progressive northward growth of the East Pacific Rise postulated by Herron. Figure 8 also shows that the existence of an extinct ridge segment between the Orozco and Clipperton Fracture Zones is permitted but not demanded by the data.

The data in Figure 8 come from two independent sources: basement ages from drill sites and magnetic anomalies. The uncertainties of the first set have been evaluated by van Andel and Bukry (1973); those of the second group are difficult to estimate. However, a recent revision by Blakely (1974) of the Miocene part of the geomagnetic polarity time scale of Heirtzler and others (1968) suggests uncertainties on the order of ±2 m.y. in the assignment of ages to all but the youngest numbered anomalies.

Consequently, the fitting of a spreading-rate curve to these two sets of data is a somewhat arbitrary procedure (van Andel and Bukry, 1973), and only a general trend that minimizes changes in rate can be established. The rates fitted in Figure 8 change from slightly more than 6 cm/yr north of the Clarion Fracture Zone to more than 8 cm/yr south of the Clipperton Fracture Zone. The rates in Figure 8 were kept as constant as possible, but other fits, including changes in spreading rate, are permitted by the data. Of such possible changes, only the acceleration around 50 m.y. B.P. indicated by the location and age of site 163 seems to be well documented (Winterer, 1973). North of the Clarion Fracture Zone, the basement age of site 41 given by van Andel and Bukry (1973) was used; if this is too young (see App. 2), the resultant spreading rate would be about 10 percent less.

Using the curves in Figure 8, the basement isochron chart of Figure 7 was constructed and used to assign basement ages to drill sites that did not reach basement and to surface cores. The distribution of basement ages between the Clarion and Clipperton Fracture Zones demands the assumption of a fracture zone approximately halfway between them. Such a fracture was predicted by Sclater and others (1971), but topographic evidence is inconclusive. The isochron pattern does not accommodate the age for site 75 of 31.5 ± 3.7 m.y. Even assuming the existence of a major fracture zone at lat 4.5° S connecting the extinct ridge south of lat 5° N with the Galapagos Rise (Rea and Malfait, 1974; Rea, 1974; Mammerickx and others, 1975), the spreading rate resulting from an initial position of site 75 on the Galapagos Rise is much too high. Perhaps the basement at this site was formed in connection with flows and intrusions from the Marquesas Islands Ridge, or perhaps true oceanic basement was not reached (Tracey and others, 1971).

A northward or northwestward movement of the Pacific plate was first postulated by Francheteau and others (1970) on the basis of paleomagnetic data from seamounts. Subsequently, northward plate drift has been postulated to explain the progressive northward displacement with increasing age of equatorial deposits in drill sites in the equatorial region (Hays and others, 1972; Tracey and others, 1971; van Andel and Heath, 1973b; Winterer, 1973). Simultaneously, Morgan (1972) and other

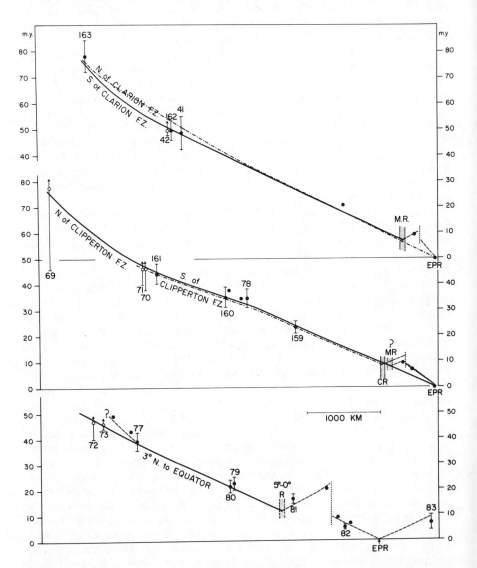

Figure 8. Relationship between basement age and distance to ridge crest. Blocks between fracture zones shown separately. Lines of best fit drawn by hand. Drill sites (black dots) are given with error bars after van Andel and Bukry (1973). Estimated ages of drill sites that did not reach basement are shown as open circles with arrows. Arrows begin at the age of oldest cored sediment; circle marks the assumed age; and the rest is error indication. Stippled vertical bars indicate crests of Mathematicians Ridge (MR), Clipperton Ridge (CR), and 5° N–0° ridge; EPR is East Pacific Rise. Vertical dotted line is boundary between crustal segments generated at two different rise crests.

authors have shown that linear volcanic trends, such as the Hawaiian, Emperor, and Line Island chains, may be interpreted as trails resulting from drift of the plate over fixed melting spots deep in the mantle. If this concept is correct, the orientation and age progression of such trails can be used to determine the absolute rotation parameters of the plate, provided that the melting spot has remained fixed in the mantle. Winterer (1973) questioned this last assumption and postulated a southward movement of the Hawaiian melting spot, but Clague and Jarrard (1973) failed to find factual support for this drift. Molnar and Atwater (1973), using reconstructions of several lithospheric plates, also found evidence of the motion of individual melting spots relative to each other, but their conclusions rest on some unproven assumptions. Recently, the fixed-melting-spot explanation for some linear island chains has been questioned (Shaw and Jackson, 1973; Schlanger and others, 1974). Therefore, the degree of coincidence of the equatorial axis of maximum sedimentation with the paleoequator that can be achieved by various rotation schemes becomes an important independent test.

The evidence from the Hawaiian chain suggests that during Cenozoic time, the Pacific plate rotated around two successive absolute poles, the rotations being represented by the Emperor and Hawaiian volcanic chains (Clague and Jarrard, 1973). Numerous other volcanic lineations in the Pacific can be fitted into the same scheme (Fig. 9). The first pole of rotation, the Emperor, has a position at approximately lat 17° N, long. 107° W (Clague and Jarrard, 1973). The Tuamotu and Marshall chains also reflect around this pole. The Line Islands, which roughly fit a great circle around the same pole, do not seem to show the required northward progression of ages (Schlanger and others, 1974). Other positions of the same pole are given at lat 23° N, long. 110° W (Morgan, 1972) and lat 23° N, long. 108° W (Winterer, 1973).

The second and most recent rotation is determined by the Hawaiian pole. Its position, based on the Hawaiian, Cook-Austral, and Guadelupe chains, has been calculated at lat 72° N, long. 83° W (Clague and Jarrard, 1973), but the circle of confidence is large and includes poles at lat 67° N, long. 73° W (Morgan, 1972) and lat 67° N, long. 45° W (Winterer, 1973). Minster and others (1974), using two independent approaches—one based on all available indicators of instantaneous plate motions, the other on all presumed hot-spot lineations—obtained a best fit for the absolute Pacific pole at lat 67° N, long. 58° W, which is also within the same confidence circle. Although this is an instantaneous solution, they suggested that it may be valid for the past 10 m.y.

Uncertainty is also associated with the determination of the rates of rotation and the time of shift from the Emperor pole to the Hawaiian pole. The data are mainly derived from the Hawaiian chain and are extrapolated from age data for the past 4 m.y., with a few supporting older data points. Shaw (1973) and Dalrymple and others (1974), discussing the problem associated with linear extrapolations of dates in a chain as complex as the Hawaiian one, concluded that such extrapolations are very hazardous. Scattered information, both biostratigraphic and isotopic, from other volcanic trends in the Pacific (Clague and Jarrard, 1973; Winterer, 1973) and from trend changes in magnetic anomalies (Morgan, 1972) has also been used in support of assumed rotation schemes.

It is therefore not surprising that numerous rates of rotation have been proposed and that the time when the change from the older to the younger pole took place is controversial. Estimates for this time vary from 27 m.y. B.P. (Clague and Jarrard, 1973) to 50 m.y. B.P. or earlier (Shaw, 1973). Clague and Jarrard also considered 42 to 44 m.y. B.P. and 37.5 m.y. B.P. as possibilities. Clague and Dalrymple (1973) and Dalrymple and others (1974) favored 42 to 44 m.y. B.P.; Morgan (1972)

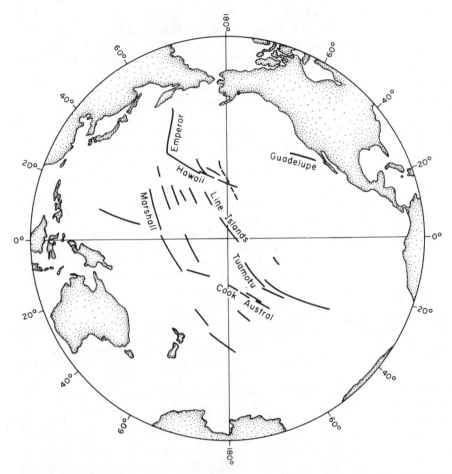

Figure 9. Volcanic lineations in the Pacific that may represent hot-spot trails. Simplified after Winterer (1973).

preferred 40 m.y. B.P., but also suggested 55 m.y. B.P.; and Jackson and others (1972) cited 25 m.y. B.P. as the most probable time within a range from 21 to 37.5 m.y. B.P. Winterer (1973), using data from other Pacific volcanic lineations as well, inferred 30 m.y. B.P., and van Andel and Heath (1973b) concluded from a change in the rate of northward shift of the axis of equatorial sedimentation that the most probable time was between 30 and 40 m.y. B.P. Data recently obtained on Leg 32 of the Deep Sea Drilling Project (Larson and others, 1973) support a time around 50 m.y. B.P.

The time of changeover from one trend to the other has a large effect on the calculated rates of rotation. Winterer (1973) gave 0.5° to 0.7°/m.y. for the Emperor pole, Morgan (1972) suggested 0.75°, and Clague and Jarrard (1973) proposed 0.8°, at least for the time prior to 42 to 44 m.y. B.P. Clague and Jarrard also suggested that between 20 and 44 m.y. B.P. rotation was either very slow (0.5°/m.y.) or insignificant and that during the past 20 m.y., rotation around the Hawaiian pole took place at 1.3°/m.y. Other proposed rates are 1.1°/m.y. (Winterer, 1973), 0.85°/m.y. (Morgan, 1972), and 0.83°/m.y. (Minster and others, 1974).

The tectonic data show that the positions of the absolute poles of rotation of the Pacific plate, the time of change from one pole to another, and the appropriate rates of rotation cannot be unambiguously defined. Independent information provided by the northward shift of the equatorial axis of maximum sedimentation is thus very important. We examine this evidence below and establish a rotational scheme for use in this study.

One drill site (site 83) has always been located on the Cocos plate, whereas another one (site 81) participated in the motion of that plate for the period 16.5 to 15 m.y. B.P. Information regarding the absolute motions of the Cocos plate is scarce and inconclusive (see review by Anderson, 1974). Larson and Chase (1970) proposed a pole at lat 40° N, long. 110° W for the Pacific-Cocos rotation; this is in close agreement with the pole computed by Minster and others (1974) at lat 41° N, long. 108° W, from which they computed an absolute Cocos pole at lat 23° N, long. 119° W. Truchan and Larson (1972) cited evidence that prior to 7 m.y. B.P., the Pacific-Cocos pole may have been located farther north near lat 78° N, long. 136° W, from where it shifted southward in response to the reorganization of the eastern edge of the Pacific plate. The migration history of this earlier pole is not well enough known to serve as a guide in determining drill-site migrations. Fortunately, the latitudinal ranges involved are small for the short life span of sites 81 and 83.

REGIONAL STRATIGRAPHY AND LITHOFACIES

Two elements of the tectonic evolution of the central Pacific Ocean have had a major influence on the depositional history of the region: (1) the subsidence of the basement with increasing age as it moved westward away from the plate edge and (2) the northward migration of the Pacific plate across the Equator.

Three long cross sections (Fig. 10) can be constructed from the drill-site data to illustrate the effect of the tectonic history on the sedimentary evolution of the region. Two east-west sections (Figs. 12 and 13) trend approximately at right angles to the accreting plate edge. The northernmost section of the two (Fig. 11) is now located about 10° of latitude north of the Equator; the other one is essentially on it. During the time of deposition of the oldest sediments, both were located considerably farther south. The Eocene section in Figure 11 was deposited very near the Equator.

Figure 10. Location chart for biostratigraphic-lithostratigraphic profiles of Figures 11 to 14. See caption of Figure 11 regarding inclusion of site 162 instead of 161.

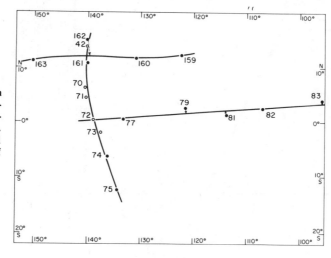

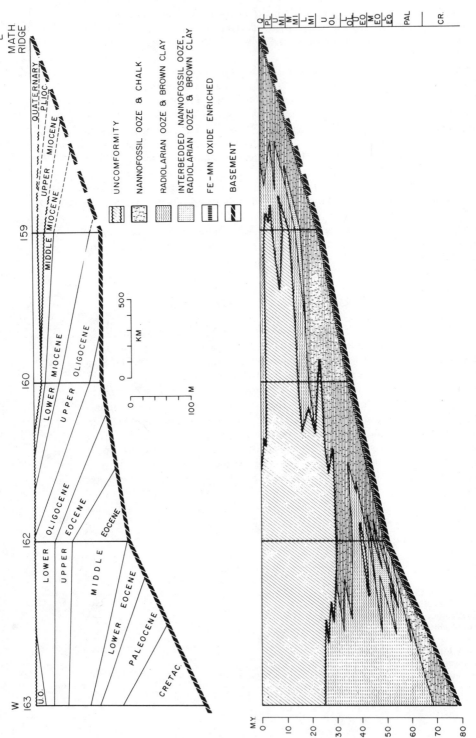

Figure 11. East-west biostratigraphic-lithostratigraphic profile north of the equatorial zone. Stratigraphic boundaries and lithology from Appendix 3. Site 162 instead of 161 was included to keep profile outside core of equatorial zone of maximum carbonate deposition; a profile including site 161 is given by van Andel and Heath (1973b, Fig. 8). Upper section is based on depth in hole; lower section is based on absolute age.

Figure 11 illustrates the diachronous nature of the lithofacies and the upward convergence of the stratigraphic boundaries expected from subsidence of the sea floor through the calcite compensation depth with increasing age (van Andel and Heath, 1973b). At most sites, basement is immediately overlain by thin sediment layers enriched in iron and manganese oxides (Cronan and others, 1972), which are common on the present East Pacific Rise crest and were precipitated from hydrothermal solutions (Dymond and others, 1973). On this basal deposit rests a thick sequence of calcareous sediments commonly grading upward from foraminiferal nannofossil ooze and chalk to radiolarian nannofossil ooze. These deposits were formed during early subsidence in moderately shallow water. Higher in the section, calcareous sediment is overlain by siliceous radiolarian ooze of decreasing carbonate content and ultimately, at the extreme western end, by pelagic brown clay. The siliceous and calcareous facies are commonly separated by a zone of interbedded calcareous and siliceous beds.

At the western end of the profile, a major unconformity represents removal of much or all of the post-Oligocene sediment. The onset of the erosional phase is difficult to determine. In part it could be a continuous or intermittent process of removing the thin deep-water sediment shortly after deposition. However, the occurrence at site 163 of a thick sequence of presumably slowly deposited radiolarian ooze, which represents more than 40 m.y. of deposition, suggests that erosion is not always a companion of slow deposition in deep water but was restricted to specific times or locations. Although it is possible that erosion took place only during Quaternary time, as postulated by Johnson (1972a, 1972b), the general decrease in age of the unconformity toward the east and the presence of several smaller unconformities at site 159 suggest that it began as early as 25 to 30 m.y. ago.

Figure 11 shows several anomalies within the general pattern. At site 162, the deposits immediately above basement are not calcareous as expected but consist instead of siliceous ooze and brown clay. This suggests a carbonate solution level shallower than that prevailing today. Above these sediments a thick sequence of calcareous Oligocene deposits is found. Because the crust must have subsided continuously, the occurrence of calcareous sediment overlying siliceous beds is anomalous. At this time, the site may have crossed the equatorial zone of maximum carbonate deposition, there may have been a large increase in the calcite compensation depth, or the equatorial zone of carbonate sedimentation may have broadened considerably.

At the Equator, the lithostratigraphic profile is quite different (Fig. 12). As before, the lithofacies boundaries are diachronous, and the stratigraphic boundaries converge upward to the west. The profile at the Equator, however, differs in that the sequence is almost entirely calcareous; siliceous deposits occur only as intercalations between calcareous beds and, in a position similar to site 162, near the base of the Eocene sequence at the western end. The interbedded calcareous-siliceous facies does not seem to result from a westward increase in depth; its thickest development lies east of site 79, and the maximum thickness occurs near the crest of the extinct ridge (Mathematicians-Clipperton ridge system). The possible lithologic effects of this fossil ridge have not been sketched in the profile because of the speculative nature of the reconstruction. Evidently, the interbedded facies is not just the equatorial equivalent of the deep-water siliceous facies farther north; a Miocene period of alternating siliceous and calcareous deposition even at shallow depth seems indicated.

At site 79 and probably also at site 72, late Eocene siliceous ooze is separated from basement only by the ubiquitous basal metalliferous zone. As at site 162, this occurrence of siliceous sediment at a shallow depositional depth suggests a

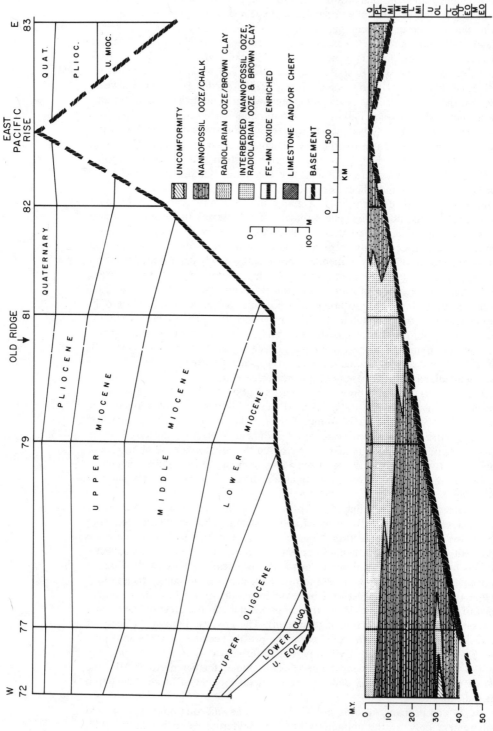

Figure 12. East-west biostratigraphic-lithostratigraphic profile in the equatorial zone. Stratigraphic boundaries and lithology from Appendix 3. Upper section is based on depth in hole; lower section is based on absolute age.

very shallow Eocene calcite compensation depth.

The equatorial profile contains no large unconformities. The sedimentation rates are much higher than farther north and south, and greater quantities of sediment must therefore be removed to create a significant hiatus. Moreover, existing hiatuses may be more difficult to detect by biostratigraphic means. A minor hiatus of Oligocene age occurs at the western end of the profile.

The north-south section (Fig. 13) is dominated by the equatorial zone of maximum sedimentation, a phenomenon that apparently has existed at least since middle Eocene time. Each of the stratigraphic intervals has a well-defined and centrally located thickness maximum, and they thin rapidly to the north and south. Each interval also has a calcareous core and grades laterally into siliceous ooze and pelagic clay. Major unconformities, which include most or all of the middle and late Cenozoic section, occur at both ends of the traverse. Commonly, they appear to have primarily affected the lateral deposits, where the rate of deposition was much smaller than in the center, but early Miocene and Oligocene calcareous sediment at the northern end of the profile has also been in part removed. Toward the Equator, the major unconformities break up into smaller units separated by zones of deposition. This indicates that the erosion was not restricted to the Quaternary but may have begun as early as middle Miocene time in the south and perhaps even earlier in the north. Another major unconformity in the extreme south, affecting the upper Eocene sediment and ranging in age from late Eocene to early Oligocene near the Equator, has no northern counterpart.

The lens shape of individual stratigraphic units and the northward shift of the axis of maximum thickness with increasing age are very evident when individual chronologic units are considered (Fig. 14). From middle Eocene time to the present, the point of maximum thickness has been displaced approximately 15° of latitude; because a displacement of the equatorial current system is improbable, plate migration must be the cause. During middle Cenozoic time, the central calcareous core of each lens was much wider, and the thickness of each unit was from two to five times greater than either before or after that time. The Eocene calcareous zone, on the other hand, was exceptionally thin and narrow. Thus, the Eocene-Oligocene boundary marks a large environmental change from a dominantly siliceous to a dominantly calcareous regime. Most probably, this was due to an increase in the rate of carbonate supply. A widening and intensified Oligocene-Miocene zone of biogenic productivity may have been a contributing factor (Chap. 7).

The equatorial zone of sedimentation can be defined more precisely on isopach maps, which use information from all drill sites, than on cross sections. The isopach maps of Figure 15, in contrast to similar maps used in Chapters 4, 5, and 6, are based exclusively on biostratigraphic correlation (Table 1) and are free from possible distortions resulting from flaws in the absolute chronology. The axis of the equatorial zone of maximum sedimentation is seen to migrate northward, reaching lat 10° N for deposits of early Oliogocene age. In general, the axis is parallel to the Equator, but its azimuth cannot be established within better than 10° to 15°. The latitudinal position of the axis also has some uncertainty; even for the Quaternary, it cannot be determined to better than ±2°. The data do not permit identification of the equatorial axis in Eocene time.

Figure 15 also shows that the southern boundary of the erosional zone north of the Equator has not varied much in latitude when the data points are migrated back in the direction of plate motion. Toward the eastern margin of the area, the isopachs tend to be deflected northward and southward where the ridge crest shoals and the rate of sedimentation increases. With increasing age, the curvature and, consequently, the ridge crest shift westward, indicating either an eastward

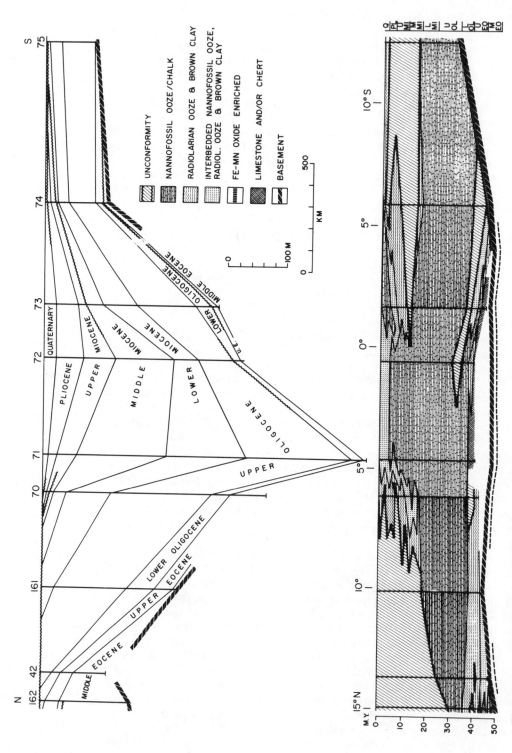

Figure 13. North-south biostratigraphic-lithostratigraphic profile approximately along long. 140° W. Stratigraphic boundaries and lithology from Appendix 3. Location shown in Figure 10. Upper section is based on depth in hole; lower section is based on absolute age.

UNCONFORMITY

NANNOFOSSIL OOZE/CHALK

RADIOLARIAN OOZE & BROWN CLAY

INTERBEDDED NANNOFOSSIL OOZE, RADIOL. OOZE & BROWN CLAY

FE-MN OXIDE ENRICHED

LIMESTONE AND/OR CHERT

BASEMENT

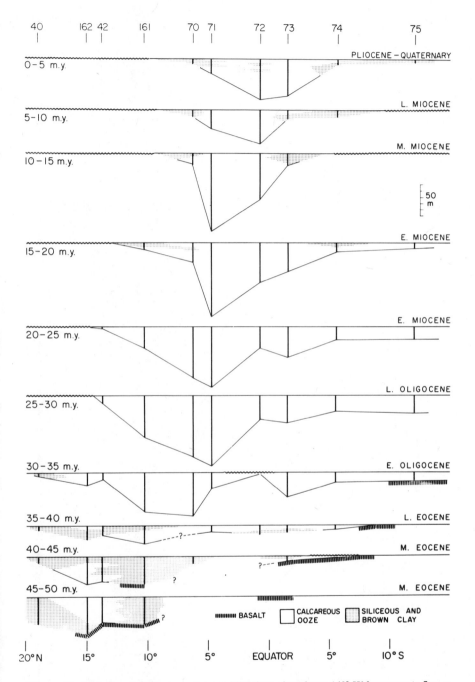

Figure 14. Variation of thickness and general lithology along long. 140° W for separate 5-m.y. intervals. Location shown in Figure 10; data from Appendix 3. Stratigraphic names are approximate. Wavy lines mark erosion; heavy dashed line is basement.

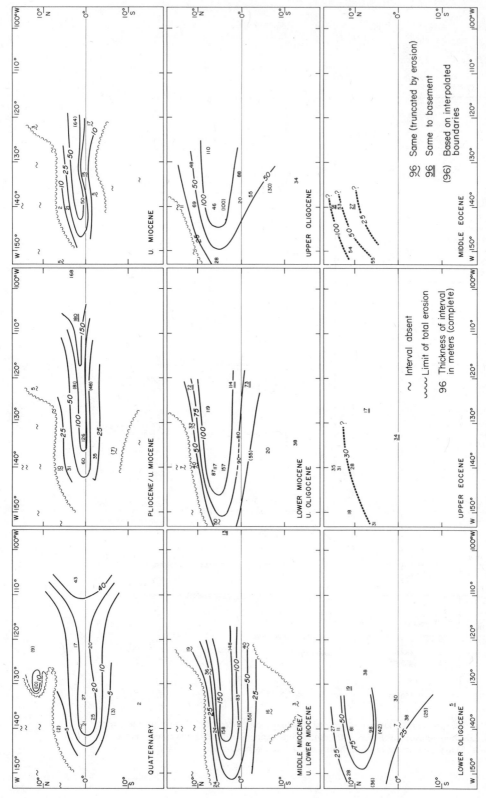

Figure 15. Isopach maps in metres for 9 time planes between the present and middle Eocene time, based on biostratigraphic boundaries defined in Table 1. Period names are approximate. Heavy dashed lines on Eocene maps indicate very uncertain isopachs. Paleolatitudes and paleolongitudes were obtained by migrating drill sites as in Figure 18.

migration of the ridge crest or a strong westerly component in the plate rotation or both.

It is evident from the foregoing that the rotation of the Pacific plate and the concomitant migration of drill sites relative to the Equator, as well as the gradual subsidence of the basement with increasing age, are dominant influences on the depositional history of the region. This pattern, however, is obviously also influenced by other significant factors. The data indicate that the calcite compensation depth has not remained constant during the past 50 m.y., but that it may have been much shallower in Oligocene time. There are also strong indications that the calcite compensation depth was very deep during Oligocene and early Miocene time and (or) the equatorial zone of sediment production was much broader and more intense. Major erosion may have started as early as the middle Miocene, both in the north and south, but was absent or much more restricted during earlier Cenozoic time. Before these problems can be addressed, however, the migration paths of the drill sites across the Equator must be determined.

MIGRATION OF DRILL SITES RELATIVE TO THE EQUATOR

The equatorial sedimentation zone is characterized primarily by its high rate of deposition and hence by the presence of a narrow, elongate zone of thick deposits parallel to and approximately on the Equator (Fig. 15). The change with time in the position of this axis measures the amount and direction of the migration of the Pacific plate and of the drill sites located on this plate. For this purpose, the equatorial axis can be defined in various ways. Some authors have assumed

TABLE 1. DEFINITION OF STRATIGRAPHIC UNITS
USED IN FIGURE 16

Approximate age (m.y.)	Name	Biostratigraphic definition
0		
	Quaternary	
2		Top *Discoaster brouweri* or top *Pterocanium prismatium* or base N22 zone
	Pliocene- uppermost Miocene	
6.5		Base N17 or top *Ommatartus antepenultimus* zone
	Upper Miocene	
11		Base *Ommatartus antepenultimus* zone
	Middle Miocene- upper lower Miocene	
18		Base *Calocycletta costata* zone
	Lower Miocene- uppermost Oligocene	
24		Base *Lychnocanium bipes* or base *Sphenolithus ciperoensis* or base P22.
	Upper Oligocene	
29.5		Base *Sphenolithus distentus* zone
	Lower Oligocene	
37.5		Base *Theocyrtis tuberosa* zone
	Upper Eocene	
44		Base *Podocyrtis chalara* zone
	Middle Eocene	
49		Top *Theocampe mongolfieri* zone

that the maximum deposition rate in the profile of any drill site marks the time
at which this site crossed the Equator (Clague and Jarrard, 1973; Winterer, 1973).
However, the rate of deposition in the central Pacific is not only a function of
latitude, but it also increases with decreasing water depth and varies with temporal
changes in the calcite compensation depth. Maxima due to these causes cannot
a priori be distinguished from those resulting from latitudinal changes. Other
investigators have identified the axis from cross sections along the long. 140° W
traverse of Figure 13 (Hays and others, 1972; Tracey and others, 1971; van Andel
and Heath, 1973b). This approach disregards nearly one-half of the available
information and suffers from a large interpolation error due to uneven spacing
of control points. Such cross sections do permit, however, some distinction between
thickness maxima resulting from different causes.

Isopach maps for a series of stratigraphic intervals, such as those of Figure
15, optimize the utilization of data and permit a clear distinction between maxima
related to latitude and maxima resulting from other causes. To be useful, however,
the intervals must be defined in terms of absolute age and be as short as possible
to increase the precision with which the position of the axis at various times
can be determined. For these reasons we will use the sedimentation-rate maps
of Figure 32 (Chap. 5) to determine the plate migration rather than those of Figure
15. These maps can be viewed as a set of isopachs for 1-m.y. intervals, spaced
in time in such a manner that changes in sedimentation rate due to effects other
than plate migration are minimized.

The successive axes of equatorial sedimentation show a gradual northward shift
with age (Fig. 16) and a small, possibly gradually increasing angle to the Equator.
Position and azimuth of the axes are subject to some uncertainty, which increases
with increasing age and decreasing number of control points. As far back as middle
Oligocene time, the azimuth error is probably near ±10°, and the distance error
about ±2° of latitude; however, for the three oldest axes, the uncertainties are
obviously greater (Fig. 32).

The poles and rates of rotation discussed in the first section of this chapter
can be used to rotate the axes of maximum sedimentation back to positions coincident
with the Equator. Of the two poles of rotation involved, the Emperor pole is
the least controversial; we have accepted a mean position at lat 17° N, long. 117° W
and a rate of rotation of 0.8°/m.y. as representative. The end of this period of
rotation is not well established; proposed times vary from 20 to more than 50
m.y. B.P. The time adopted here of 50 m.y. B.P. is supported by recent drilling
on Koko Guyot near the Emperor-Hawaii bend (Larson and others, 1973) and
minimizes the number of drill sites that are affected. In order to obtain the correct
rotations, the Emperor pole itself must first be rotated back to its proper position
around the Hawaiian pole.

Poles proposed for the Hawaiian rotation are much more scattered. For trial
purposes we may use lat 72° N, long. 83° W (Clague and Jarrard, 1973) and lat
67° N, long. 59° W (Minster and others, 1974) as representative of the cluster,
with a rotation rate of 0.83°/m.y. (Minster and others, 1974). The results of these
rotations are shown in Figure 17. The rotation around lat 72° N, long. 83° W clearly
results in an excessive inclination relative to the Equator and a fairly large scatter,
especially for the Oligocene and Eocene axes. Rotation around lat 67° N, long.
59° W produces a much-reduced inclination to the Equator, which is within the
uncertainty limits of the azimuth estimate but is perhaps indicative of a pole position
somewhat to the east of the one adopted. For the period 0 to 25 m.y. B.P.,
the axes cluster tightly on the Equator, but for the older ones the adopted rate
of rotation is clearly too large.

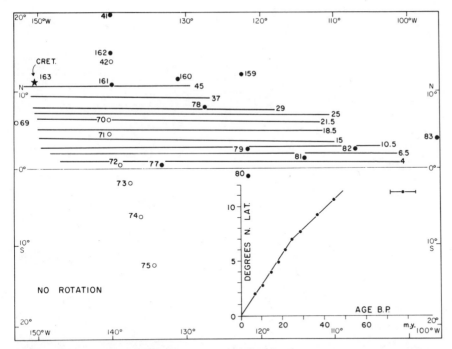

Figure 16. Northward shift of the axis of maximum sedimentation with increasing age. Axes from Figure 32 are migrated to present position and represent sedimentation maxima from isopach maps for 1-m.y. intervals ±1 m.y. High later Cretaceous rate at site 163 shown with star. Numbers at right of axes are midpoints of isopach intervals in millions of years. Numbered dots are drill sites. Insert: latitude of axes plotted against age; curve fitted by hand. Note change in rate of shift at approximately 25 to 30 m.y. B.P. Point with error bar in upper right corner of graph is rate maximum of site 163.

Much better clustering can be obtained by assuming a reduced rate of 0.25°/m.y. for the period 25 to 50 m.y. B.P. (Fig. 18). The only exception is the 45-m.y. axis, which is, however, very poorly defined because of the sparse control points that are located near the rise crest in shallow water. Better clustering could also be obtained for the rotation around lat 72° N, long. 83° W by a change in rate around 25 m.y. B.P.; this would require an increase by 50 to 75 percent in the later rate, which is an unreasonably large figure. A change in the rate of rotation around the Hawaii pole at 20 to 30 m.y. B.P. has been proposed by Clague and Jarrard (1973) on independent grounds and by van Andel and Heath (1973b). It is supported by a recently obtained age of approximately 30 m.y. B.P. for the sea floor near Midway Island (Larson and others, 1973).

The pole and rate of rotation can be obtained directly from Figure 16 by deriving a polar-wandering track relative to the pole of rotation of the Pacific plate and computing absolute pole and rate of rotation from this track. The errors involved in this procedure are large, primarily because of its sensitivity to the azimuth of the axes, which is poorly defined, and because of uncertainties in time. The pole so obtained lies at lat 64° N, long. 55° W, and the rotation rate is 0.8° to 0.9°/m.y. for the interval between 0 and 25 m.y. B.P. and 0.2° to 0.3°/m.y. for the earlier period. The pole contains within its error circle both poles used in Figure 17.

The validity of the rotational parameters for the Emperor pole cannot be ascertained with drill-site data. The reasonable fit obtained in Figure 18 indicates that the

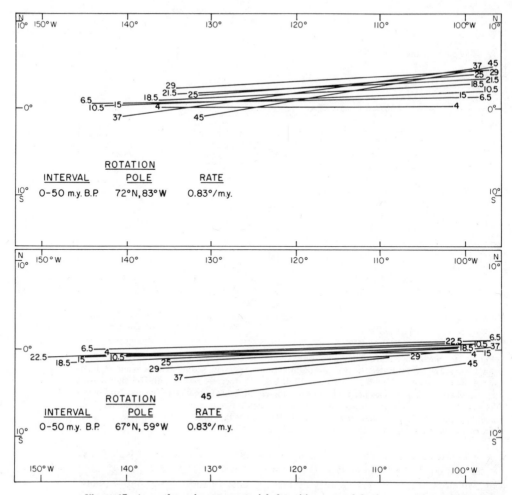

Figure 17. Axes of maximum equatorial deposition rotated back to position in relation to Equator at time of deposition. Top: rotation around Hawaii pole of Clague and Jarrard (1973). Bottom: rotation around Hawaii pole of Minster and others (1974). Axes from Figure 16, identified with age in millions of years.

assumed time of change from the Emperor to the Hawaii pole at 50 m.y. B.P. is either approximately correct or no later than 40 to 45 m.y. B.P. Site 163 has two sedimentation-rate maxima. The latest one, at 43 m.y. B.P., is used in the determination of the 45-m.y. axis. The other one, at 75 m.y. B.P., lies close to the sediment-basement contact and cannot be placed on the Equator with any reasonable rotation scheme. It must be attributed to shallow deposition on the rise crest.

Two drill sites have been located on the Cocos plate for all or part of their existence. Because of the short times involved (16.5 to 15 m.y. B.P. for site 81 and 8.2 m.y. B.P. for site 83), their migration paths are short. We have accepted a Cocos pole at lat 23° N, long. 119° W and a rotation rate of 1.47°/m.y. (Minster and others, 1974) for the reconstruction of these paths. It is probable that the Cocos plate has had a more complex history, but the data are inadequate for a reliable reconstruction. The two sites have remained close to the Equator for their entire histories and provide no independent evidence.

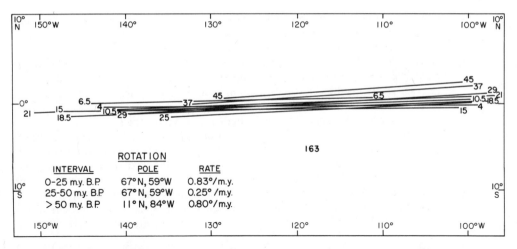

Figure 18. Adjusted rotation of axes of maximum equatorial deposition according to scheme adopted for this study. Rotation prior to 50 m.y. B.P. around Emperor pole adjusted for adopted Hawaiian rotation. Axes from Figure 16 identified with age in millions of years. Late Cretaceous (75 m.y. B.P.) maximum at site 163 indicated with site number.

The drill-site migration paths computed with the Pacific plate rotation scheme of Figure 18 and from the assumed pole and rotation rate for the Cocos plate are shown in Figure 19. Paleolatitudes and paleolongitudes computed from these paths at 1-m.y. intervals are listed in Appendix 3.

BATHYMETRIC EVOLUTION

When rotated back to their initial positions, the drill sites define the plate edge at the time of formation of the basaltic basement. Figure 20 shows a reconstruction of the eastern edge of the Pacific plate for several periods for which data are available. Between 50 and 20 to 30 m.y. B.P., the eastern edge of the Pacific plate apparently migrated rapidly eastward from long. 115° W to 105° W. This migration is a consequence of the fact that the spreading rate for this period was larger than the small westward component, which resulted from early slow rotation around the Hawaiian pole. At 25 m.y. B.P., the rotation rate on the Hawaiian pole was tripled, approximately balancing the spreading rate, and the plate edge has maintained its position since that time. The jump from the Mathematicians-Clipperton-5° N-0° system to the present East Pacific Rise around 10 to 15 m.y. B.P. did not permanently change this stationary character. The coincidence in time between the major reorganization of the Pacific plate edge around 20 to 30 m.y. ago, the change in the rate of rotation, and establishment of a stationary plate edge may indicate a genetic relationship.

The Cenozoic paleobathymetric evolution of the region can be reconstructed in some detail from the paleodepths and paleopositions of the drill sites, provided that the validity of the rotational procedures and of the basement age to basement depth relationship is accepted. For the younger stages, the present bathymetry was used as a guide, but with increasing age, the lack of control forced greater generalization. In general, the change of topography with time is not great and mainly reflects a gradual change in orientation resulting from the rotation of the plate. A particularly interesting example of this is the shift in position of the broad shallow area that now lies south of lat 10° S and west of the East Pacific

Rise. The origin of this broad shoal, which also has an anomalously low heat flow, is in doubt; it could be crust riding over an asthenospheric bulge (Menard, 1973), but the gravity data do not support such an explanation (Mammerickx and others, 1975). This broad shoal influences the present sediment distribution by extending the zone of calcareous deposits far south of where it normally would be (Fig. 1). Its middle Cenozoic position, according to Figure 21, was significantly farther north. As will be shown later, sedimentation rates and carbonate distribution patterns support this part of the paleobathymetric reconstruction.

There are several important second-order deviations from the present topography that have implications for the tectonic and depositional history of the region. The first of these is a gradual change in the depth of the crest of the ancestral East Pacific Rise (Fig. 22). The present crestal depth is approximately 3,000 m. From this value, the mean depth changed to about 2,700 to 2,800 m for the period between 40 and 45 m.y. B.P. and then increased to 3,000 m in Late Cretaceous time. Given the many assumptions that enter into estimates of past crestal depths

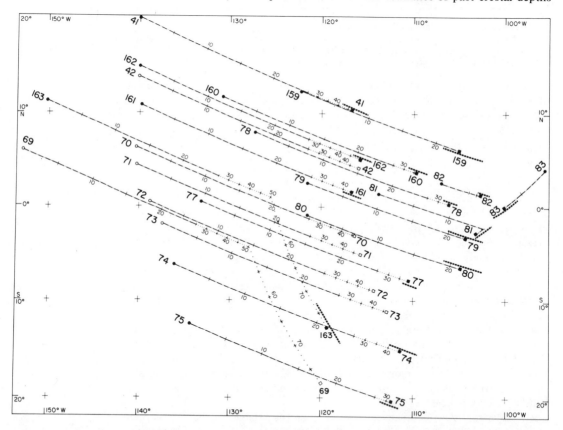

Figure 19. Migration tracks of drill sites used in this study. Site 83 rotated around Cocos pole at lat 23° N, long. 119° W at 1.47°/m.y.; same rotation for site 81 during interval 16.5 to 15 m.y. B.P. Later track of site 81 and all other tracks according to scheme of Figure 18. Black circles are present positions; black squares are initial positions of sites that penetrate basement. Open circles and open squares represent present and initial positions, respectively, of sites terminating in sediment. Tracks marked in 1-m.y. steps (small dots); crosses mark 5- and 10-m.y. points starting with 0 B.P. at present position. Dashed bars near initial site positions indicate range of basement age uncertainty (after van Andel and Bukry, 1973). Rotation of sites 69 and 163 prior to 50 m.y. B.P. around Emperor pole at lat 11° N, long. 84° W. This pole was obtained by migrating the present Emperor pole back to its position at 50 m.y. B.P. according to scheme of Figure 18.

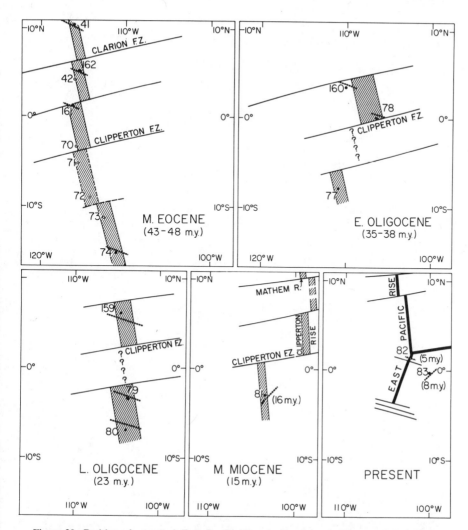

Figure 20. Position of ancestral East Pacific Rise during the past 50 m.y. B.P. Obtained by rotating major fracture zones and drill sites back to appropriate positions and sketching plate edge through initial drill-site positions. Uncertainty of initial drill-site positions shown with error bars. For middle Miocene map, entire Mathematicians–Clipperton–5°N–0° system was rotated back to site 81 initial position. Black dots, sites where basement was penetrated; open circles, sites ending in sediment. Basement ages of sites noted only on diagram for present configuration. Rotation scheme from Figure 18.

and the highly indented nature of present bathymetric contours, the rather large scatter does not justify tectonic inferences. On the other hand, the two southernmost sites (74 and 75), which are now located on the above-mentioned broad shoal, had very shallow crestal depths of less than 2,500 m at about 40 m.y. B.P. In view of the large extent of the shoal, these crestal depths may be meaningful.

In the course of the past 50 m.y., the ridge has gradually widened from a rather narrow feature in late Eocene time to the broad one of today. This progressive widening, which has been accompanied by a marked decrease in the steepness of the upper slope, is illustrated in Figure 23. The present East Pacific Rise is strongly asymmetric, with an east flank resembling the slope of the west flank

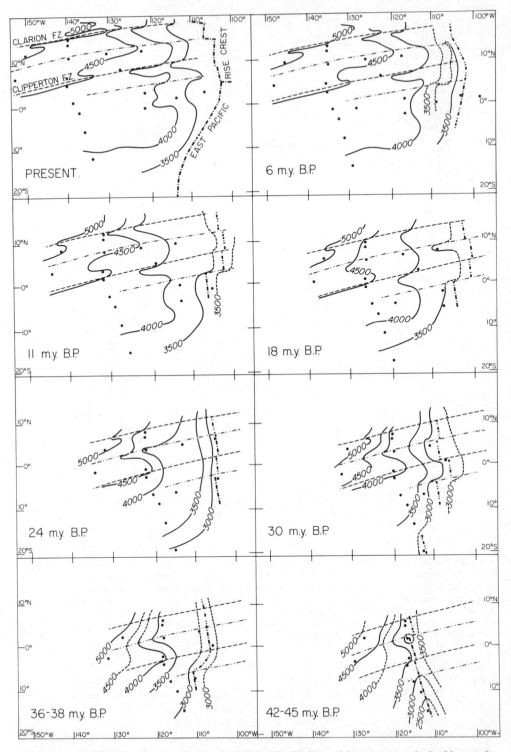

Figure 21. Paleobathymetry of the central equatorial Pacific. Positions of fracture zones obtained by rotation according to scheme of Figure 18; plate edge determined by interpolation from Figure 20. Depth contours in metres based on paleobathymetric data from Appendix 3. Present bathymetric contours from Chase and others (1970) and Mammerickx and others (1975). Black dots are drill-site locations. Time levels correspond to Figure 15 but are based on absolute ages from Appendix 3 rather than on purely biostratigraphic correlations of Table 1.

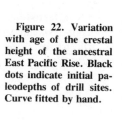

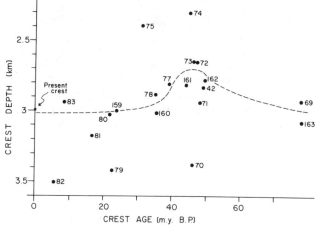

Figure 22. Variation with age of the crestal height of the ancestral East Pacific Rise. Black dots indicate initial paleodepths of drill sites. Curve fitted by hand.

as it was 25 m.y. ago (Mammerickx and others, 1975). Thus, during early and middle Cenozoic time, the rise may have been rather symmetrical, the present asymmetry perhaps being induced by the southeastward migration of the broad shoal now located south of lat 10° S.

Vogt and others (1969) suggested a relationship between the width and elevation of the central part of a spreading ridge and the spreading rate. Sclater and others (1971) presented an empirical and qualitative analysis showing that the crests of slow spreading ridges (less than 3 cm/yr) may be 200 to 500 m shallower than those of fast spreading ridges. They were unable to establish whether slow spreading ridges subside more rapidly with increasing age. If this were so, the subsidence curves would eventually merge, and the slow spreading ridges would be shallower, narrower, and steeper and have a smaller cross-sectional volume than fast spreading ridges, with concomitant consequences for the volume of the ocean basins (Berger and Winterer, 1974; Hays and Pitman, 1973). This suggests that the East Pacific Rise may have spread more slowly during early Cenozoic time than during either the latter part of the Cretaceous or late Cenozoic time. The information on spreading rates needed to test this inference is sparse. Larson and Pitman (1972, Fig. 8) presented evidence for a period of fast spreading during the latter part of the Cretaceous, but their data do not indicate a middle Cenozoic change in spreading rate. Berger and Winterer (1974) and Baldwin and others (1974) have pointed out, however, that the absolute chronology for the latter part of the Cretaceous is very uncertain and that the postulated fast spreading might be an artifact. Winterer

Figure 23. Change in profile with time of the west slope of the ancestral East Pacific Rise. Profiles constructed from paleobathymetric maps of Figure 20 for area just north of Clipperton Fracture Zone except for slope labeled "south" which is located at about lat 10° S. Numbers indicate age (in m.y. B.P.).

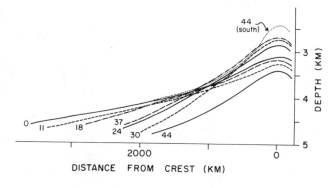

(1973) compiled data to show the existence of a period of slow spreading before 40 m.y. B.P., but he saw no evidence for a period of fast spreading in later Cretaceous time. The question must thus remain open. Whatever the causes of the inferred changes in the profile of the East Pacific Rise may be, they and the migration of the broad southern shoal may have significantly affected the depositional patterns by changing subsidence rates, by raising the crest into shallower water with less carbonate dissolution, and by narrowing the profile to speed the transition into deep water with little or no deposition of calcium carbonate.

4

Calcium Carbonate Distribution and the History of the Calcite Compensation Depth

It is evident that large changes in the nature and areal distribution of equatorial Pacific pelagic sediments would accompany a rise or fall of the calcite compensation depth (CCD) (Chap. 2). Lithostratigraphic evidence is given in Chapter 3 showing that the CCD has indeed varied its position with time, as was shown earlier by Heath (1969a) on the basis of surface cores. Berger (1973a) and van Andel and Moore (1974) presented curves for the change in depth of the CCD during Cenozoic time based on DSDP data, taking into account plate migration and sea-floor subsidence.

Berger (1968, 1970a, 1973b) and Moore and others (1973) have shown that carbonate particles are strongly dissolved at depths much shallower than the CCD. Moore and others (1973) suggested that the rate of dissolution increases most rapidly in the upper few kilometres of the water column below the supersaturated surface water. Berger (1968) defined a zone (the lysocline) where 50 to 80 percent dissolution of foraminiferal fauna occurs. In the central Pacific, the lysocline appears to coincide with the abyssal thermocline at the top of the northward-flowing Antarctic Bottom Water and, also, with a level at which Peterson (1966) found a marked change in the rate of dissolution of calcite spheres. Despite the observed changes in foraminiferal fauna, however, the lysocline does not correspond to a significant change in the carbonate content of sediment deposited at this depth. This is because 50 percent dissolution of a sediment consisting of 96 percent carbonate still leaves a deposit containing 92 percent carbonate. Such carbonate values are among the highest of those occurring in DSDP cores. The example illustrates the insensitivity of the carbonate content to abrupt changes in dissolution rates associated with water-mass boundaries.

Heath and Culberson (1970) and Berger (1971) have shown that the CCD is an inevitable consequence of increasing dissolution with depth (Fig. 24). Consequently, its presence does not automatically imply a change in water-mass properties. This is apparent if one considers the calcite rate of accumulation rather than the calcite content of the sediment. The former appears to decrease uniformly and perhaps in a linear fashion below the lysocline (Heath and Culberson, 1970), whereas the latter, expressed as (carbonate accumulation rate/total accumulation rate) × 100, changes little when the two rates are comparable, but it decreases rapidly as the carbonate accumulation rate approaches zero. This is reflected by the strongly bimodal distribution of carbonate percentages in all the samples studied: 26 percent

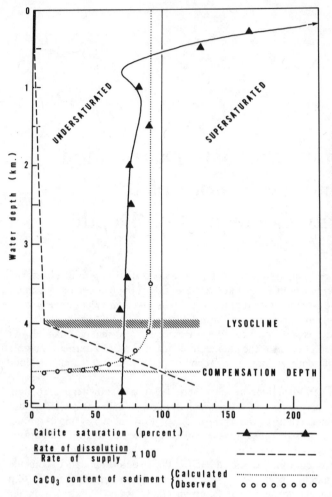

Figure 24. Parameters influencing the distribution of calcium carbonate in equatorial Pacific sediment (after Heath and Culberson, 1970, Fig. 3). Observed carbonate content after Bramlette (1961); rate of dissolution from Peterson (1966). Saturation curve from Li and others (1969) for the central Pacific, recalculated with apparent dissociation constants of Culberson and Pytkowicz (1968) and apparent solubility product of Hawley and Pytkowicz (1969).

have less than 25 percent $CaCO_3$ and 45 percent have more than 75 percent. Consequently, the depth of the CCD and the sharpness of the transition between calcareous and noncalcareous deposits is a function of three more fundamental variables: the depth of the abyssal thermocline (closely approximated by the lysocline), the rate of increase of the dissolution rate with depth, and the rate of supply of carbonate and noncarbonate detritus to the sediment.

Because carbonate percentages are readily available and carbonate accumulation rates are not, it has been the general practice to infer spatial and temporal changes in carbonate dissolution from changes of the CCD rather than from the more fundamental variables just named. In drawing such inferences, the nonlinear relationship between changes in these variables and changes in the CCD should be kept in mind.

Figure 25 gives the distribution of calcium carbonate in the central equatorial Pacific for 11 intervals during the past 50 m.y. that correspond to the intervals shown on the maps of deposition rates discussed in Chapter 5. The most pronounced feature of Figure 25 is the well-developed calcareous zone along the Equator, especially during the past 12 m.y., when the zone has been narrow, centered symmetrically on the Equator, and bordered by steep gradients. Prior to this time, the zone was generally broader, less well defined, and bordered by gentler gradients, particularly during middle and early Miocene time. Maximum carbonate values in the center of the zone generally exceed 75 percent, ranging from 75 to 95 percent. For the intervals between the 6 and 7 m.y. B.P. and 10 and 11 m.y. B.P., the central maximum is less extensive, and carbonate values are lower. In Table 2, the width of the equatorial zone and the mean high carbonate content are summarized. The most notable feature is the increase in width of the equatorial zone and the occurrence of maximum carbonate values during Oligocene and early Miocene time.

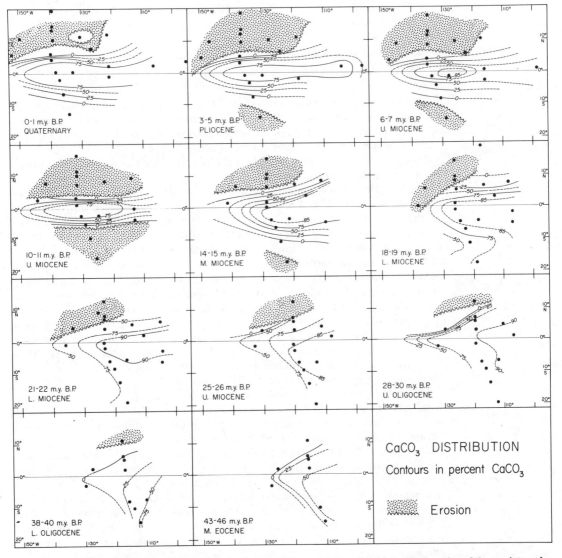

Figure 25. Isopleths of calcium carbonate concentration in the central equatorial Pacific for selected 1-m.y. intervals. Contours in percentage of total sediment are based on average concentration for all 1-m.y. units in the interval at each drill site (black dots). Mapped intervals are identified on Figure 31 (shaded bars on right). Drill sites rotated according to scheme of Figure 18.

TABLE 2. WIDTH AND MEAN HIGH CARBONATE CONTENT
OF THE EQUATORIAL ZONE

Interval (m.y. B.P.)	Width* (km)	CaCO$_3$† (%)
0–1	350	81
3–5	275	78
6–7	175	81
10–11	150	78
14–15	225	82
18–19	>500	87
21–22	>500	89
25–26	300	85
28–30	275	87
38–40	..	..
43–46	225	..

*Northern half-width in kilometres at approximately long. 130° W.
†Mean of values within 75 percent contour.

At the end of the Eocene the carbonate content of equatorial sediments increased dramatically from the low values of early Cenozoic time to the much higher ones of the Oligocene and later. The Eocene equatorial zone is rather poorly defined, with maximum carbonate values clustering around 50 percent.

Prior to 15 m.y. B.P., the symmetry of the equatorial carbonate zone was distorted by a large area of anomalously high carbonate values in the southeast part of the region. The paleobathymetric reconstruction (Fig. 22) shows that this anomaly coincided with a broad and relatively shallow region that extended far to the west of the rise crest. At present, a similar shallow region lies south of the study area, where it is covered with highly calcareous deposits (Fig. 4); however, during early and middle Cenozoic time, it apparently extended north of lat 10° S.

The bilateral symmetry about the Equator is well demonstrated by a plot of carbonate isopleths on a paleolatitude-age coordinate system (Fig. 26). This diagram clearly illustrates the dramatic change in carbonate content at the end of the Eocene, the symmetry of the equatorial zone, and the sharp gradients on either side of it (Berger, 1973a). On the other hand, the view that the lithology of the post-Eocene deposits in the equatorial Pacific can be predicted from present depth, latitude, basement age, and basement depth alone is clearly oversimplified, as is shown by comparison of the carbonate isopleths with the paleodepths on the same paleolatitude-age coordinate system (Fig. 26, bottom).

The gradual decline in the width and carbonate content of the equatorial zone from early Oligocene to Pliocene time (Fig. 26) led to Pliocene conditions that almost duplicated those of middle and late Eocene time. This severity of the Pliocene oceanic environment has been inferred previously from faunal assemblages and preservation states. Erosion apparently is not closely related to the carbonate content of the underlying and overlying deposits (Fig. 25). The younger unconformities as well as the oldest ones are associated with sediments of low carbonate content that represent slow deposition. Thus, the unconformities may represent periods of nondeposition as well as erosion. On the other hand, the removal of thick sections of the highly calcareous northern equatorial sediments of Miocene and Eocene age indicate persistent, vigorous, and concentrated activity of bottom currents.

At the present time, the equatorial CCD is depressed by about 500 m (Berger and Winterer, 1974) relative to its level in parts of the Pacific farther to the north and south; this is due to the large equatorial input of carbonate particles

(Heath and Culberson, 1970). Berger (1973a) has shown that a depressed equatorial CCD can be traced back to at least middle Eocene time. In Figure 27, the depth variation of the CCD across the Equator is shown for two intervals. These intervals were chosen because the width and composition of the equatorial carbonate-rich sediments remained stable long enough to furnish adequate data (Figs. 25, 26; Table 2). A value of 20 percent $CaCO_3$ was chosen as the boundary between calcareous and noncalcareous sediment. This value is the minimum in the bimodal frequency

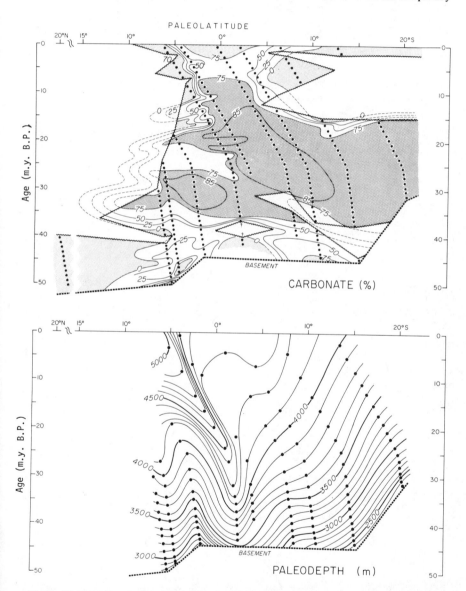

Figure 26. Isopleths of carbonate concentration in percentage of total sediment plotted in an age-paleolatitude diagram for drill sites on north-south traverse. Top: carbonate concentration isopleths. Bottom: paleodepth contours. Black dots are data points; heavy dashed line at bottom of each diagram represents basement; wavy line indicates erosion surface. From left to right: sites 41 (not on lower graph), 162, 42, 161, 70, 71, 72, 73, 74, and 75.

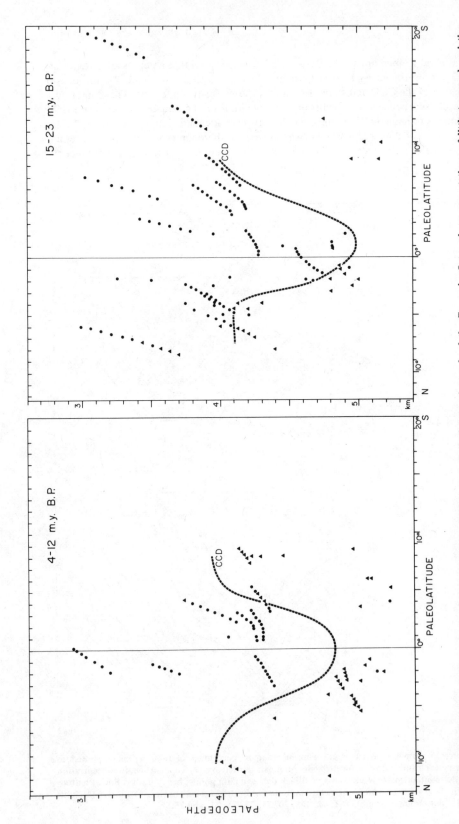

Figure 27. Variation of calcite compensation depth (CCD) with paleolatitude for two intervals of the Cenozoic. Intervals represent times of little temporal variation of CCD. Triangles: less than 20 percent $CaCO_3$; dots: more than 20 percent (Apps. 3, 5).

distribution of all carbonate analyses. Other investigations have used lower values, but fewer than 20 of the 130 data points are involved, and the precise limit has no influence on the position of the curves of Figures 27 and 28. Although the data are limited, the depression of the equatorial CCD by about 800 m over a zone of 6° to 8° of latitude is evident. Consequently, the equatorial CCD must be distinguished from its normal Pacific counterpart when tracing the history of carbonate dissolution in the equatorial region of the Pacific. Berger's (1973a) data also suggest that the two CCD's may have varied separately, at least in part, during the past 50 m.y.

Although the data are somewhat unevenly distributed, the positions of the two CCD curves are reasonably well defined, and the differences between them are obvious (Fig. 28). Least well defined is the CCD for the Pacific during Oligocene time, because the curve merely defines the lower limit of the available data. There is independent evidence for this position, however, from inferences based on carbonate accumulation rates (Chap. 7). On the other hand, van Andel and Moore (1974) placed a much deeper Oligocene CCD for the Pacific very close to the equatorial CCD. The paleolatitudes of their data points were obtained from a rotation scheme (Fig. 17, bottom) that assumed fast northward migration during the period 50 to 25 m.y. B.P. The scheme of Figure 19 produces a much better fit of the equatorial zone of maximum sedimentation to the Equator, but it eliminates a number of data points from the set used for the determination of the Oligocene CCD for the Pacific in particular from site 69.

During the Eocene, the Pacific CCD (Fig. 29) was close to 3,200 m, 500 m shallower than previously assumed (Berger, 1973a). Very early in Oligocene time, at 37 ± 2 m.y. B.P., it dropped rapidly to at least 4,300 m. During latest Oligocene time, it began a gradual rise to a brief minimum of 3,800 or 3,900 m in the late Miocene. For its present position, Berger and Winterer (1974, Fig. 1) suggested 4,500 m at lat 5° N and 5° S, but any depth between 4,000 and 4,500 m is possible (Fig. 28). This uncertainty is not surprising in view of the complex history of carbonate dissolution since the beginning of the late Cenozoic global glaciations (Arrhenius, 1952; Hays and others, 1969).

The Eocene CCD for the equatorial region is not well defined but was probably not much below 3,400 m. Near the Oligocene-Eocene boundary it dropped simultaneously with the Pacific CCD, but to a much deeper level near 5,100 m. Around 20 m.y. B.P., the equatorial CCD rose rather steeply to a minimum at about 4,800 m and then more gradually to 4,700 m or shallower (broken curve) in late Miocene time. Its position during late Pliocene and Quaternary time is well defined by surface core data of Hays and others (1969). The present position is somewhat uncertain; Bramlette (1961) and Heath (1969a) gave 4,700 m, whereas Berger and Winterer (1974) preferred 5,000 m. Our data are most compatible with 4,900 m or a little more.

Heath's (1969a) curve was based on the present depths of his samples from surface cores. If his depths are corrected for subsidence, his curve becomes quite similar to ours. Berger's curve, which was derived by procedures similar to our own and, except for surface cores, was based on the same data, is nearly identical to the equatorial CCD except for a much deeper level in the Eocene. This difference is the result of our acceptance of carbonate-free samples in sites 42 and 162 as indicators of the position of the CCD. This interpretation might be questioned, because in several Pacific DSDP sites, carbonate-free intervals very near basement (Cronan and others, 1972) resulted from rapid precipitation of metalliferous sediment of hydrothermal origin (Dymond and others, 1973). The sediments in question, however, do not resemble such hydrothermal deposits, and they occcur well above

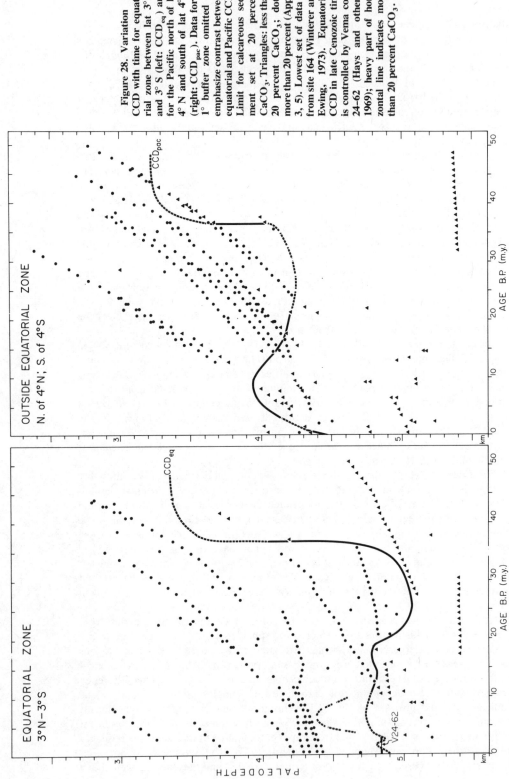

Figure 28. Variation of CCD with time for equatorial zone between lat 3° N and 3° S (left: CCD_{eq}) and for the Pacific north of lat 4° N and south of lat 4° S (right: CCD_{pac}). Data for a 1° buffer zone omitted to emphasize contrast between equatorial and Pacific CCD. Limit for calcareous sediment set at 20 percent $CaCO_3$. Triangles: less than 20 percent $CaCO_3$; dots: more than 20 percent (Apps. 3, 5). Lowest set of data is from site 164 (Winterer and Ewing, 1973). Equatorial CCD in late Cenozoic time is controlled by Vema core 24-62 (Hays and others, 1969); heavy part of horizontal line indicates more than 20 percent $CaCO_3$.

basement (25 m or more, corresponding to an age difference of 5 to 10 m.y.).
Evidence from other oceans supports the existence of a very shallow CCD during
Paleocene and Eocene time (Berger, 1972; Pimm, 1974).

Temporal variations in carbonate dissolution, which cause the changes in depth
of the CCD, cannot be inferred from the position of the CCD alone (Heath and
Culberson, 1970). Therefore, we must examine the temporal changes in the carbonate
accumulation rates (Chap. 5).

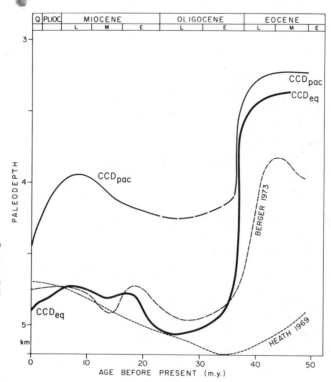

Figure 29. Comparison of Pacific CCD (CCD$_{pac}$) and equatorial CCD (CCD$_{eq}$) from Figure 28 with CCD curves of Heath (1969a) and Berger (1973a).

5

Cenozoic Rates of Deposition

CONTROLLING FACTORS

Biogenic calcareous and siliceous material strongly dominates sediments of the central equatorial Pacific Ocean. Only a small fraction, probably variable in size with time (Heath, 1969b), consists of volcanic ash, authigenic minerals, and land-derived components. This fraction becomes important only where the rate of deposition of biogenic material is very low. In the eastern equatorial region, however, the Neogene equatorial deposits have a higher clay content, which increases toward the east (Hays and others, 1972). The biogenic material is produced almost entirely near the surface of the sea, and the benthic component is minor. Insufficient data exist to estimate the regional variation of production rates of this material in the central equatorial Pacific. Consequently, production rates of biogenic material and their variation in space and time must be inferred from variations in the rate of deposition.

Two main factors control the rate of deposition: regional variations in productivity and the solution of calcium carbonate at depth. Siliceous components are also strongly dissolved both in the water column and after deposition, but there is no evidence that silica dissolution is a function of depth, and the calcite compensation depth has no opal equivalent (Heath, 1974). Consequently, depth changes should be reflected in the rate of deposition of calcium carbonate and also, because carbonate is commonly a dominant constituent, in the bulk deposition rate, but not in the rate of deposition of biogenic silica. Changes in productivity in the surface waters, on the other hand, should affect the deposition rates of both the calcareous and siliceous components.

In the following discussion we shall use the term "rate of deposition" as a general one and reserve "rate of sedimentation" for its measurement in metres per million years and "rate of accumulation" for grams per square centimetre per thousand years (App. 1).

SEDIMENTATION RATES

Because increasing dissolution of carbonate with depth and the productivity maximum at the equator are dominant factors, the rate of sedimentation at each drill site should be maximal just above basement and again at the time the site crossed the equatorial zone. In Table 3, pertinent data have been compiled; examples are shown in Figure 30. Eleven sites cross the Equator, six have distinct maxima, one shows a shoulder on a larger peak, two have no maximum, and one has

49

a hiatus at this time. Of the thirteen sites that reached basement, eight show a maximum, three do not, and two have a maximum for the accumulation rate but not one for the sedimentation rate. The heights of the maxima vary considerably but are commonly well above adjacent mean values. The ages of the equatorial maxima vary by only 1 or 2 m.y. from the computed time of Equator crossing; this variation is in good agreement with the estimated age uncertainty. The basal maxima commonly occur 1 or 2 m.y. after initiation of deposition. This may be a result of initially slow deposition of metalliferous sediment directly on basement (Dymond and others, 1973) or, more likely, of an initial phase of alternating erosion and deposition high on the rise crest (van Andel, 1972). The basal maximum at site 78 is 5 m.y. younger than the oldest sediment; it is doubtful that this can be regarded as a true shallow-water maximum.

Together, these sedimentation rate maxima constitute only a small fraction of the total. Of the 45 pronounced maxima at 20 drill sites, only about one-third can be attributed to an Equator crossing or a shallow basal paleodepth. When all the sedimentation-rate curves of all drill sites are superimposed (Fig. 31), the

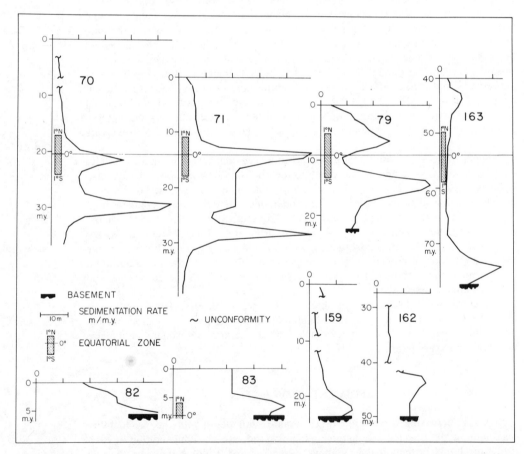

Figure 30. Sedimentation-rate changes with time for selected drill sites. Top row: drill sites that cross the Equator (2°-wide zone of crossing indicated with shaded bar). Note absence of equatorial maximum at sites 79 and 163. Bottom row: sedimentation-rate change near sediment-basement interface. Sedimentation-rate scale at lower left.

TABLE 3. SEDIMENTATION RATE MAXIMA AT EQUATOR CROSSINGS
AND IN BASAL SEQUENCE OF DRILL SITES

Site	Equator crossing				Contact age (m.y.)	Basal sequence		
	Time of crossing (m.y. B.P.)	Average rate* (m/m.y.)	Peak age (m.y.)	Peak rate (m/m.y.)		Average rate† (m/m.y.)	Peak age (m.y.)	Peak rate (m/m.y.)
42		(not crossed)			(not reached)			
69	18.5	hiatus			(not reached)			
70	20.5	12.2	21.5	27.0	(not reached)			
71	14.5	25.0	13.5	50.0	(not reached)			
72	1.5	12.8	shoulder		(not reached)			
73		(not crossed)			(not reached)			
74		(not crossed)			no data			
75		(not crossed)			31.5	7.3	30.5	10.0
77	1.5	14.2	2.5	17.0	39.0	4.3	39.0	6.0
78	35.0	14.3	30.0	20.0	35.0	15.7	30.5	20.0
79	9.0	16.4	7.0	26.0	22.5	15.7	none	none
80	0.0	8.0	none		21.7	14.5	none	none
81	5.5	no data			16.5	32.9	15.5	50.0
82		(not crossed)			5.6	37.0	5.6	83.0
83	8.2	37.8	7.5	42.0	8.2	29.0	7.5	42.0
159		(not crossed)			23.5	10.6	22.5	16.0
160		(not crossed)			35.5	3.1	none	none
161		(not crossed)			44.0	7.1	44.0	14.0
162		(not crossed)			49.5	14.1	none§	none§
163	54.0	3.8	none		78.0	15.0	75.0	24.0

*Average rate at Equator crossing calculated over ±1° of latitude.
†Average rate in basal sequence calculated over nearest 500-m paleodepth change.
§Maximum present in bulk accumulation rate.

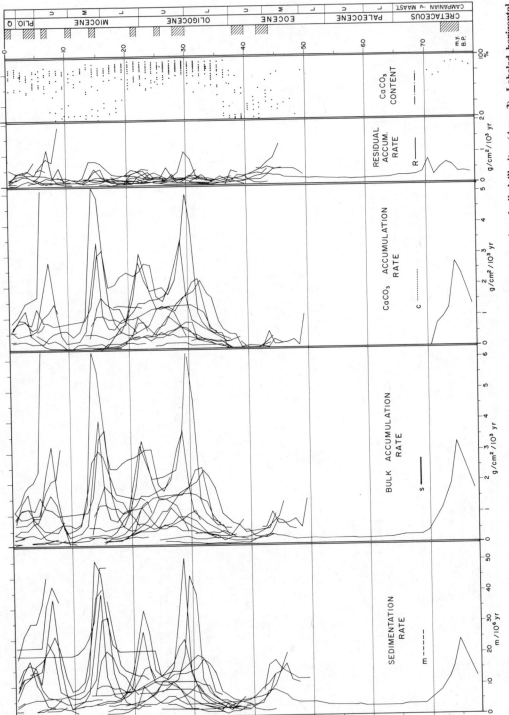

Figure 31. Variation with time of rates of deposition and of calcium carbonate content; composite of all drill sites (App. 3). Labeled horizontal lines near bottom of each column identify corresponding curves in Appendix 3. Zero values for calcium carbonate are not shown. Shaded bars on right represent intervals selected for mapping in Figures 32 and 36 through 38. Time scale from Berggren (1972a).

existence of a number of periods of rapid sedimentation is apparent. During each phase of maximum deposition, sites located far from the equatorial zone or in very deep water (App. 3) show a small peak in the sedimentation rate, whereas those near the Equator show a large peak. During intervening periods of slow deposition, even sites near the Equator show greatly reduced maxima. The individual maxima or minima of each period correlate well; the spread of ages is about 1 to 2 m.y. for the younger phases and 3.5 m.y. for the older phases, values that are within the range of the uncertainty of age assignment.

Six maxima at 2, 6 to 7, 14 to 15, 21 to 22, 28 to 30, and 42 to 46 m.y. B.P., and six minima at 0 to 1, 3 to 5, 10 to 11, 18 to 19, 25 to 26, and 38 to 40 m.y. B.P. can be distinguished. Of these, the maximum at 2 m.y. B.P. is suggested by only two distinct peaks and by three shoulders on larger peaks. A large maximum occurs at site 163 near 75 m.y. B.P., but it is not supported by other data. The spacing between maxima varies from 2 to 3 m.y. in late Cenozoic time to about 7 m.y. in early Cenozoic time.

Not unexpectedly, the various accumulation rates (Fig. 31) show the same pattern with equally good correlations, although the range in the carbonate-free accumulation rate is much smaller than that of the others. The calcium carbonate content, on the other hand, does not show the same pattern. Between 0 and 8 m.y. B.P., and again between 20 and 37 m.y. B.P., the carbonate content is high, and its range of variation is limited. In the intervening period, the scatter of values is much greater, although most fall in the range of 75 to 100 percent. Before 37 m.y. B.P., high values are rare, and most deposits (all between 50 and 70 m.y. B.P.) contain no carbonate at all. At site 163, a zone of calcareous deposits corresponding to the rate maximum occurs between 70 and 78 m.y. B.P. Since this section was deposited in relatively shallow water (above 3,600 m), this may not signify a regional increase in carbonate deposition. The large change in carbonate content around 37 m.y. B.P. can be attributed with certainty to the change in CCD discussed in Chapter 4; the reasons for the other variations are less clear.

Because of these temporal variations, isopleth maps depicting the regional pattern of sedimentation rates should be limited to narrow time intervals keyed to the maxima and minima. For the construction of such maps, the intervals designated by shaded bars on the right of Figure 31 have been selected. These include five maxima and five minima, as well as the interval from 0 to 1 m.y. B.P., for comparison with present conditions.

The principal feature of all sedimentation-rate isopleth maps (Fig. 32) is a well-defined west-trending equatorial zone of rapid deposition. This zone can be seen even in the Eocene map, notwithstanding the limited control and clustering of data points in shallow water near the crest of the mid-ocean rise. The alternation of times of minimal and maximal deposition is evident from the number of contours, but each pattern is simple, coherent, and remarkably free of anomalies.

With increasing age, the sedimentation rates tend to decrease. This is in part due to compaction, but the effect is obscured by the large decrease in carbonate content below the Oligocene-Eocene boundary. Thus, the decrease due to compaction cannot be separated from that due to the drop in carbonate content. During intervals of slow sedimentation, the rate in the equatorial zone varies from 10 to 20 m/m.y.; during times of rapid deposition this increases to 20 to 30 m/m.y. Occasionally, much higher values occur; these are restricted to the eastern part of the region, in particular the equatorial zone (for example, the intervals from 6 to 7 and 14 to 15 m.y. B.P.), and occur only at times of rapid deposition.

The width of the equatorial zone also varies. Occasionally, this seems to be correlated with the sedimentation rate: the minima at 10 to 11 and 18 to 19 m.y.

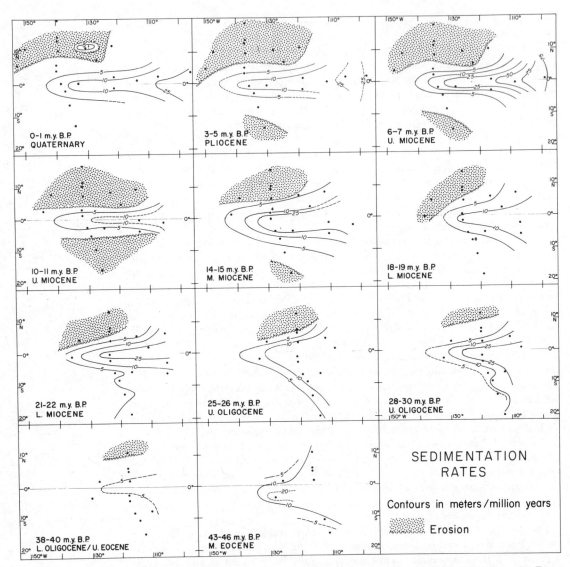

Figure 32. Isopleth maps of sedimentation rates for various intervals of the Cenozoic. Zones are from Figure 31 (shaded bars on right). Black dots are drill sites rotated according to scheme of Figure 18. Rate data (small numbers near dots) shown only where they do not fit isopleths. To contour intervals of rapid and slow deposition, respectively, highest and lowest values at each site were used. The axes of maximum equatorial sedimentation of this figure are used in Figures 16, 17, and 18.

B.P. have a much narrower equatorial zone than the intervening maxima. There is, however, a general pattern to the variation in width of the equatorial zone, which is wide between 15 and 30 m.y. B.P. and much narrower in the preceding and following periods. The considerable width during middle Eocene time is an artifact resulting from the location of most drill sites in shallow water near the crest of the rise.

During the past 15 m.y., the pattern has been highly symmetrical along the Equator. Previous to this time the symmetry was distorted by a southeastern lobe of anomalously high sedimentation rates, although the equatorial zone of high sedimentation rates has always been present. The anomalous area coincides with the large shoal zone shown in Figure 21 and with a zone of anomalously high carbonate content (Fig. 25). The patterns of sedimentation rates thus confirm the paleobathymetric evidence for a southward migration of this shoal during later Cenozoic time.

VARIATION IN DEPOSITION RATES AND BIOSTRATIGRAPHIC CONTROL

The question must be raised whether the variation of the rates of deposition could be an artifact of the differentiation of the age-depth curves (App. 1) or, more importantly, whether it could be seriously affected by distortions in Berggren's (1972a) time scale.

If we assume that this scale is valid, the uncertainty associated with the estimate of depth in the hole for each 1-m.y. time boundary increases from ±1 m.y. for the past 10 m.y. to ±3 m.y. at 50 m.y. B.P. (App. 1). This uncertainty is, however, constrained by the uncertainty of the points on either side, so that the net effect on the depth and sediment-thickness values is reduced. Because all determinations of maxima and minima involve 3 to 5 points, most maxima might be reduced, but not eliminated, and most minima might be raised by errors in the depth-age relationship.

The effect of possible systematic distortions in the Berggren time scale is potentially much larger, because uncertainties of the scale result from the small number of isotopic age determinations, the sometimes involved correlations between isotopic dates and biostratigraphic zones, and the confidence limits of the isotopic ages themselves. At the present time, it is not possible to quantify the confidence limits of the time scale, but comparison of various recent scales indicates that large internal distortions in the scale, especially for Miocene time, are possible (Moore, 1972). If such distortions exist, they would have a large effect on the estimates of rates of deposition and might be responsible for some or all of the fluctuations in Figure 31.

These questions can be approached in several independent ways. First, curves like those of Figure 31 could be determined for oceanic regions that are not part of the equatorial Pacific current system. If maxima and minima occurred at times different from those of Figure 31, the fluctuations in the central equatorial Pacific are possibly real. On the other hand, if synchronous variations are found, two conclusions would be possible: (1) the deposition of biogenic material might have varied in response to truly global causes, or (2) the fluctuations were produced by a distorted time scale. We have not yet attempted such a comparison.

Second, we can examine the plausibility of a time scale based on a constant sedimentation rate at each site except for variations due to crossing of the Equator or to deposition in shallow water just above basement. Unfortunately, plausibility is difficult to define, and instead of using a constant sedimentation rate, one might prefer a random variation produced by local erosion and redeposition. Although the latter is a much more common process than is usually envisaged (Moore and others, 1973; van Andel, 1973), there is no good evidence that local reworking has a major long-term effect on average sedimentation rates (Berger and von Rad, 1972).

In Figure 33, the biostratigraphic zones of site 77, selected because it does not have a pronounced equatorial maximum, have been plotted on a new time scale based on a constant rate of deposition. The resulting distortion of the Berggren

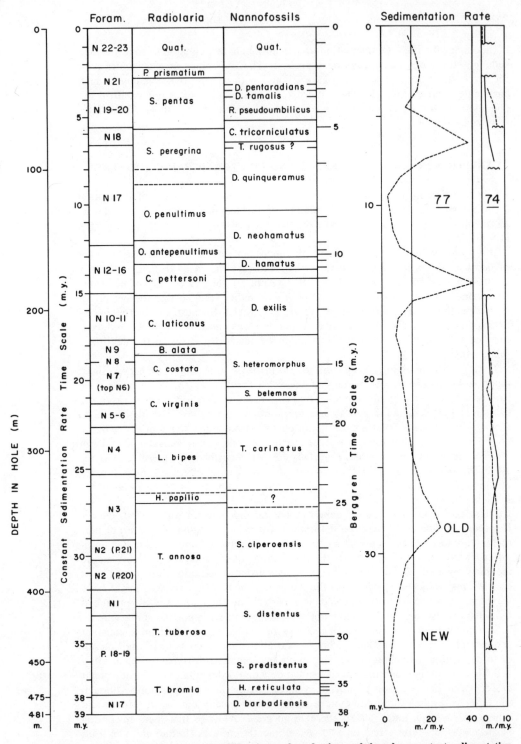

Figure 33. Comparison between Berggren (1972a) time scale and a time scale based on constant sedimentation rate for site 77. New time scale (second bar from left) obtained by dividing total depth to top of basal sedimentation peak (at 475 m) into 38 1-m.y. intervals. Biostratigraphic zones (center block) are plotted against depth in hole, yielding new ages for zone boundaries. Berggren (1972a) scale on right of center block. Right: sedimentation-rate variation according to Berggren (broken line) and new time scale (solid line). Extreme right: sedimentation-rate variation for site 74 according to Berggren time scale (broken line) and new site 77 time scale (solid line).

time scale is major and affects nearly all zone boundaries. On the extreme right, the sedimentation-rate curve for site 74, selected because it shows only small fluctuations on the Berggren scale, has been replotted on our new time scale. As might be expected, use of this new scale has merely replaced one set of fluctuations with another. In Figure 34, constant-sedimentation-rate time scales calculated separately for five drill sites are compared with the Berggren scale. The sites are distributed over the region, were selected to avoid large equatorial maxima, and are truncated above the basal maximum. The diagram suggests that the assumption of a constant sedimentation rate over any long time interval will not provide a reasonable time scale.

The Berggren time scale, used here in its latest version (Berggren, 1972a), which includes many new radiometric age determinations, is the result of a thorough treatment of marine biostratigraphies and their relationship to radiometric ages (Berggren, 1969a, 1969b, 1971, 1972a). Despite relatively sparse isotopic confirmation, we consider its application in this study justified. However, one area of

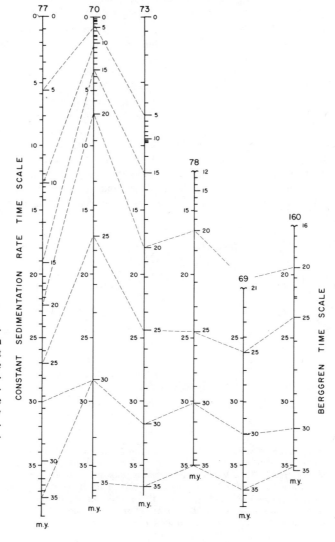

Figure 34. Comparison between Berggren (1972a) time scale (right of each bar) and time scales obtained by converting each site to constant sedimentation rate (left of each bar). New scales obtained as described in legend of Figure 33.

disagreement could markedly affect the shape of the sedimentation- and accumulation-rate curves presented here. We propose, therefore, an alternative interpretation of the biostratigraphic data.

The middle part of the Miocene is one of the few intervals of the Cenozoic for which there is an embarassment in time-stratigraphic information. The embarassment lies in the difficulty of the correlations and in the resulting disagreements (Moore, 1972). Berggren (1972a) and Berggren and Van Couvering (1974) have presented several alternatives to the time scale for the middle part of the Miocene, most of which involve some variation in the durations of the foraminiferal zones N14 through N16 (Blow, 1969). The limited scope of these alternatives is in part due to Berggren's distrust of K-Ar dates from volcanic ash and his convictions concerning the relative and absolute ages of the *Globigerina nepenthes* foraminiferal datum and the Hipparion mammalian datum. Preferring to accept only those radiometric ages that come directly from marine sections (Dymond, 1966; Rodda and others, 1967; Turner, 1970; Page and McDougall, 1970), we present in Figure 35 an alternative based on the biostratigraphic correlations presented by Berggren (1972a). The modification uses Berggren's age estimates for the base of zones N9 and N16 and radiometric dates shown in the diagram, and it is much closer to the scale found in Berggren and Van Couvering (1974) than to the earlier one (Berggren, 1972a).

In the discussion of this chapter we will use the ages based on the original Berggren (1972a) time scale as reported in Appendixes 3 and 4, keeping in mind that uncertainties in zonal boundary ages may modify the shape, height, and position of the rate maxima and minima. In the discussion on paleoceanographic history, however, we shall also take into account conversions based on the alternate time scale presented in Figure 35. The effect of this time scale is illustrated in Figure 36. The maximum at 14 to 15 m.y. B.P. is reduced markedly and slightly displaced to 14 to 16 m.y. B.P. Between 10 and 11 m.y. B.P., a new brief maximum may be in part due to a poor fit at the boundary of the two scales, but it is certain to remain even after adjustment. New minima exist at 9 and 12 to 13 m.y. B.P.

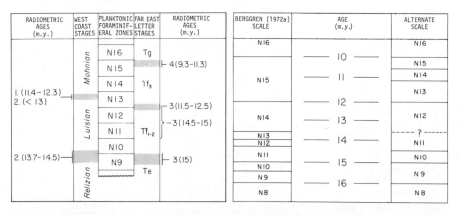

Figure 35. Radiometric data for biostratigraphic zones between 10 and 16 m.y. B.P. (left) and proposed new time scale (right). Correlations with the Blow (1969) zonation from a summary of Berggren (1972a). Relationship between proposed new time scale for middle part of Miocene and zonation of Blow (1969) is based entirely on dates in left diagram. Base of zones N9 and N16 is identical with Berggren's (1972a) scale. In left diagram, left column: 1, after Dymond (1966); 2, after Turner (1970). Center column: after Blow (1969). Right column: 3, after Page and McDougall (1970); 4, after Rodda and others (1967).

The existence of several alternations of rapid and slow deposition, however, remains clear. The converted rate data are in Appendix 1 (Table 6).

We emphasize that the chronology of the middle part of the Miocene is certain to remain in flux for some time and that we do not presume to have presented a final scale in Figure 31. In the meantime, we prefer this particular alternative, but the conversion method given in Appendix 1 can be used for conversion of the same data to other, more precise time scales as they become available.

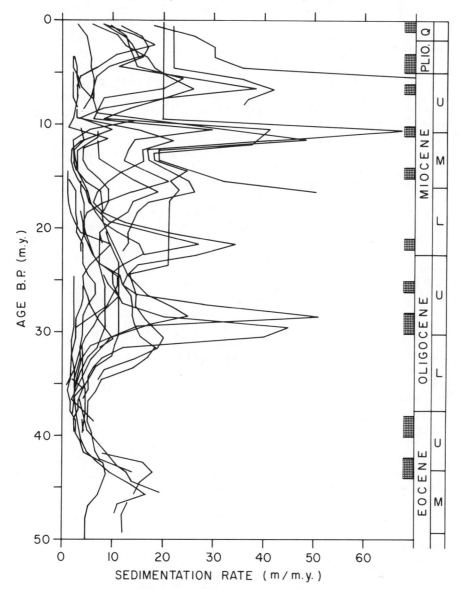

Figure 36. Variation with time of sedimentation rates in all drill sites, modified from Figure 31 with alternate time scale of Figure 35, for middle part of Miocene. Shaded bars on right are intervals of slow and rapid deposition from Figure 31. Sedimentation-rate curves of all drill sites in Appendix 3 were modified according to new time scale (App. 1, Table 6) for period of 10 to 16 m.y. B.P. and superimposed.

ACCUMULATION OF CALCAREOUS AND NONCALCAREOUS SEDIMENTS

Accumulation rates (in $g/cm^2/1,000$ yr), unlike sedimentation rates (in $m/1,000,000$ yr), are not affected by increasing age and overburden, and they permit, to a first approximation, quantitative comparisons between deposits of different ages and lithologic characteristics. Moreover, separate rates for the carbonate and carbonate-free fractions can be obtained. On the other hand, data coverage is more restricted, and the cumulative errors are somewhat larger (App. 1).

The isopleth patterns of the bulk accumulation rate (Fig. 37) are quite similar to those of the sedimentation rates, with some significant exceptions. The maps show that the gradual decrease in sedimentation rate with increasing age is an artifact and that for all periods of maximum deposition, the average rates do not differ by more than a factor of two. A drastic change in rate at the Eocene-Oligocene boundary is apparent in both data sets, but the accumulation rates indicate that Eocene deposition was not as insignificant as Figure 32 implies. Finally, the equatorial zone of maximum deposition is narrower on sedimentation-rate maps than on Figure 37, presumably because of the large effect of compaction on the less calcareous and initially much more porous deposits at greater distances from the Equator.

Separate isopleth maps of the carbonate and carbonate-free accumulation rate (Figs. 38, 39) show a well-defined and symmetric equatorial maximum since at least middle Eocene time. The carbonate maps show the familiar pattern of a large increase after the interval between 38 and 40 m.y. B.P., but the change is proportionally much greater than for the bulk accumulation rate. The carbonate-free (residual) accumulation rate, on the other hand, varies little with age and shows no large change across the Oligocene-Eocene boundary. The only real maximum occurs during the interval between 14 and 15 m.y. and has mean high values well above $0.5 \ g/cm^2/1,000$ yr. Otherwise, mean high values range from 0.2 to $0.3 \ g/cm^2/1,000$ yr for periods of slow deposition to 0.3 to $0.5 \ g/cm^2/1,000$ yr for times of rapid deposition. Accumulation rates of carbonate-free material are much higher within the equatorial zone.

For the past 8 m.y., accumulation rates of both carbonate and carbonate-free material have been very high at the eastern end of the equatorial zone. Carbonate rates there were also high during the period 28 to 46 m.y. B.P., when most drill sites were located in relatively shallow water near the rise crest, but the carbonate-free fraction does not show corresponding values.

The equatorial zone as defined by carbonate accumulation rates differs significantly from the one defined by carbonate-free rates. The carbonate equatorial zone is narrower and is bordered by closely set contours, whereas the carbonate-free equatorial zone is broad with gentle gradients. The carbonate equatorial zone is bordered by zero contours not far from the Equator, whereas appreciable carbonate-free accumulation rates occur as far north and south as data are available. The effect of the shoal region in the southeast (Fig. 21), which is evident on carbonate-content (Fig. 25) and sedimentation-rate (Fig. 32) maps, can be seen on all maps of carbonate accumulation rates before 15 m.y. B.P. and, although less distinctly, on maps of carbonate-free accumulation rates before 25 m.y. B.P.

It is apparent from Figures 32, 37, 38, and 39 that sedimentation- and accumulation-rate patterns are closely correlated with latitude. In addition, there are less well defined east-west gradients. For the bulk and carbonate accumulation rates, the east-west gradient may be due to longitudinal variation in productivity or to the effect of shoaling toward the crest of the mid-ocean rise. For the carbonate-free accumulation rate, paleodepth plays no role, but east-west changes in production or in the influx of nonbiogenic material from the east are possible.

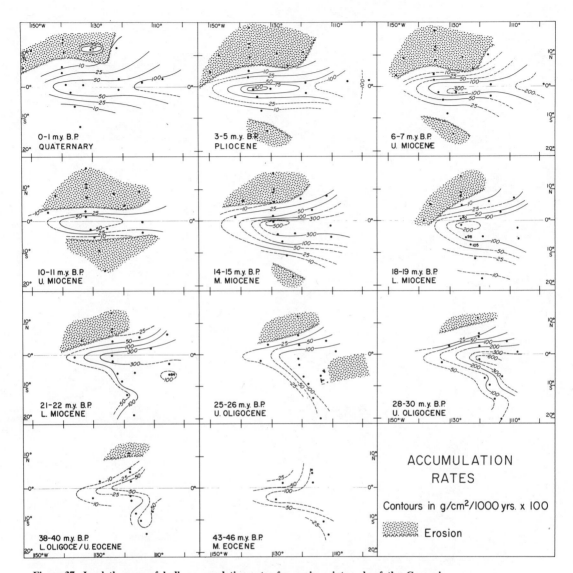

Figure 37. Isopleth maps of bulk accumulation rates for various intervals of the Cenozoic. Zones are from Figure 31 (shaded bars on right). Black dots are drill sites rotated according to scheme of Figure 18. Rate data (small numbers near dots) are shown only where they do not fit isopleths. To contour intervals of rapid and slow deposition, respectively, highest and lowest values at each site were used.

For the six youngest time planes, the distribution of the carbonate accumulation rate with latitude (Fig. 40) is quite symmetric about the Equator and terminates at fairly well defined points of zero accumulation. The width of the carbonate zone has changed little during the past 20 to 25 m.y., but peak heights vary considerably, and very low values mark the two youngest intervals. For the past 30 m.y., the peaks are centered between the Equator and a few degrees south, but the earlier ones are located somewhat farther north, perhaps owing to a slight underestimate of the rate of plate rotation. For Oligocene time the central peak is unusually broad and reaches a maximum height at that time. The southern limbs

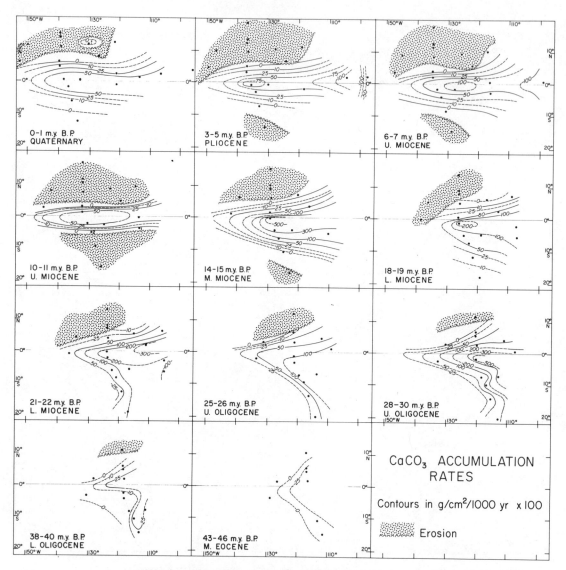

Figure 38. Isopleth maps of carbonate accumulation rates for various intervals of the Cenozoic. See legend for Figure 37 for explanation.

of the curves for the period prior to 20 m.y. B.P. are strongly asymmetric, even though a central peak is always present. This asymmetry does not appear to be simply related to paleodepth, as is indicated by the depth values on the graphs.

Plots of the relationship between carbonate-free accumulation rate and paleolatitude (Fig. 41) are also symmetric, bell-shaped distributions that are centered on the Equator. They are much broader and lower than the carbonate curves, however, and vary little in height and width with time. Exceptions are the 0 to 1- and 14 to 15-m.y. B.P. intervals, which have anomalously low and high maxima, respectively. The data for the interval from 25 to 26 m.y. B.P. do not permit the fitting of a curve, and the 43 to 36-m.y. B.P. curve is poorly defined. An

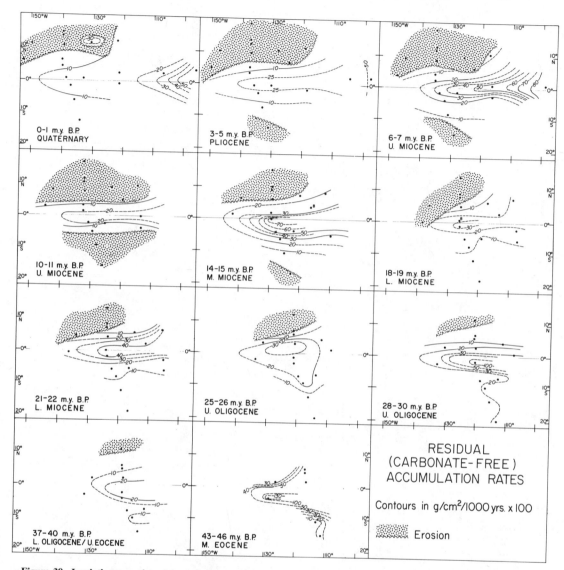

Figure 39. Isopleth maps of carbonate-free (residual) accumulation rate for various intervals of the Cenozoic. See legend for Figure 37 for explanation.

asymmetric southern limb is evident in all curves prior to 20 m.y. B.P. This anomaly is not pronounced, but it is puzzling because it cannot be attributed to shallow depth. Influx of terrigenous clay or volcanic ash from the east might be invoked as a reason, but in the absence of quantitative data regarding the composition of the carbonate-free fraction, this would be mere speculation. All peaks except for the interval from 43 to 46 m.y. B.P. are centered on the Equator and confirm the adopted rotation scheme.

Peak heights, peak widths, and the width between points of zero accumulation for the carbonate accumulation rate can be used to summarize the Cenozoic evolution of the equatorial zone of maximum deposition (Table 4; Fig. 42). Because some

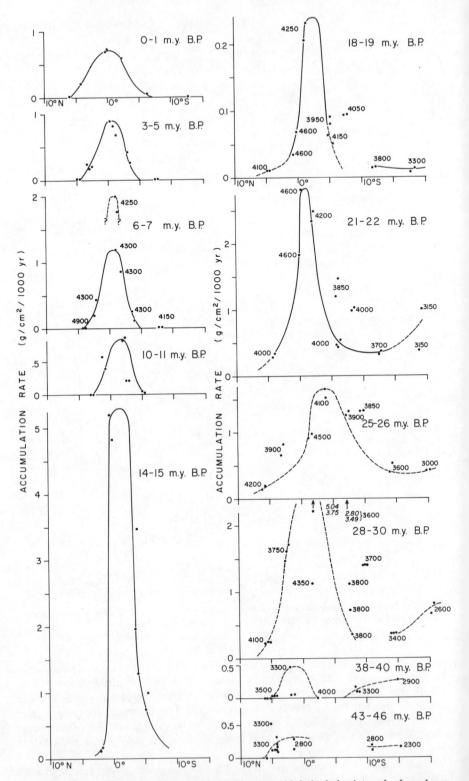

Figure 40. Variation of carbonate accumulation rates with latitude for intervals of maximum and minimum deposition of Figure 31. Based on drill sites 42, 70 to 75, 77, 161, and 162. All data points in each interval are used. Small numbers near points indicate paleodepth in metres rounded to nearest 50 m. Curves fitted by hand. Off-scale numbers in diagram for 28 to 30 m.y. (in italics) indicate high values in g/cm²/1,000 yr from sites 70 and 77.

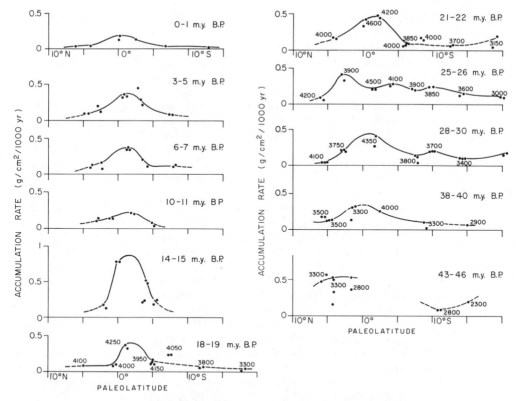

Figure 41. Variation of carbonate-free (residual) accumulation rates with latitude for intervals of maximum and minimum deposition from Figure 31. Based on drill sites 42, 70 to 75, 77, 161, and 162. All data points in each interval are used. Small numbers near points indicate paleodepths in metres rounded to the nearest 50 m. Curves fitted by hand.

peaks are markedly asymmetric, the widths were obtained by doubling the northern half-width. The equatorial zone as defined by the carbonate-free accumulation rate is essentially constant in peak values. Zone widths, on the other hand, vary in a major way. The interval from 14 to 15 m.y. B.P. is marked by a very high peak which, however, is not reflected by a corresponding increase in the width of the carbonate zone. Use of the modified 10 to 16-m.y. B.P. time scale of Figure 35 eliminates this peak (dashed curve) and creates a smaller one at 10 to 12 m.y. B.P. Thus modified, the carbonate accumulation rate at the center of the equatorial zone changes from a very low rate during the Eocene to a very high one in the Oligocene, then tapers gradually to a low value for the past few million years. Superimposed on this pattern is a series of maxima and minima that are synchronous with those of Figure 31. The width of the carbonate deposition zone behaves similarly, changing from narrow to very wide at the Eocene-Oligocene boundary, then gradually decreasing to a minimum in late Miocene and Pliocene time. The present width is somewhat greater than that of the Pliocene. The effect of the modified time scale on the width of the carbonate zone is insignificant and is not shown in Figure 42.

The curves of Figure 41, especially when modified with the alternate time scale of Figure 35, correspond closely to temporal variations of the CCD. The relationships between carbonate accumulation rates, carbonate dissolution, and carbonate supply will be discussed in more detail in Chapter 7.

VARIATION OF DEPOSITION RATES WITH DEPTH AND LONGITUDE

Carbonate content and carbonate accumulation rate are functions of the depth of the depositional surface. In the central equatorial Pacific, this surface slopes westward from the crest of the East Pacific Rise, and longitudinal changes in carbonate content and accumulation rate may occur as a result. Moreover, the traverse of drill sites at long. 140° W, on which much of the discussion above is based, increases in depth northward and has subsided continuously since middle Eocene time. Most of the increase in depth took place during Eocene and early Oligocene time, when the southern end of the traverse was particularly shallow. At all times, differences in depth between adjacent sites have been significant (Fig. 26, bottom). Consequently, paleodepth changes may well have contributed to the variations shown in Figures 40 and 42.

The relationship between depth and accumulation rate in individual drill sites is obscured by temporal fluctuations and by maxima occurring at the time of Equator crossings. The former, which were short lived, can be eliminated by curve fitting. The residence time in the equatorial zone, however, is on the order of 5 to 10 m.y., and its effect is major; thus, only sites located entirely within or outside the equatorial zone should be used. Their number is small and includes only one equatorial site (site 83). The relationship between depth and accumulation rate (Fig. 43) is, as expected, nonlinear, with rates increasing toward shallower depth. The curves are similar in shape for different sites, but the rates for any given depth vary greatly with the site. Theoretically, an approximate doubling of the rate of carbonate accumulation for each 300 m of shoaling at the shallow end could be expected (Berger, 1971), but this occurs only rarely. This may be due to major winnowing and downslope dispersal of sediment near the rise crest as observed by Moore and others (1973) and van Andel (1973).

The relationship between carbonate accumulation rate and depositional depth

TABLE 4. VARIATION IN DIMENSIONS OF THE EQUATORIAL ZONE OF
MAXIMUM DEPOSITION DURING THE PAST 50 M.Y.

Interval (m.y. B.P.)	Carbonate width (km)	Accumulation zero width (km)	Rate height (g/cm²/1,000 yr)	Carbonate-free width (km)	Accumulation rate height (g/cm²/1,000 yr)
0–1	900	1,500	0.70	600	0.17
3–5	650	1,050	0.90	500	0.35
6–7	550	1,000	1.20	500	0.35
8–9	··	··	(0.78)	··	(0.23)
10–11	600	900	0.85 (2.55)	550	0.22 (0.65)
12–13	··	··	(1.93)	··	(0.31)
14–15	500	1,100	5.20	500	0.85
15–16	··	··	(2.00)	··	(0.49)
18–19	550	1,500	2.40	450	0.40
20–21	600	1,800	2.80	650	0.45
25–26	1,400	2,400	1.80	?	?
28–30	1,400	2,200	3.75	600	0.40
38–40	600	1,000	0.45	650	0.35
43–46	?	?	0.30?	?	0.50

Note: Data from Figures 40 and 41. Width estimates are based on northern half of equatorial zone. Width of carbonate accumulation rates measured at one-third of peak height; width of carbonate-free rate measured at two-thirds of peak height. Data in parentheses are from conversion to alternate time scale for middle of Miocene (Fig. 35). Zero width is width at points of zero carbonate accumulation.

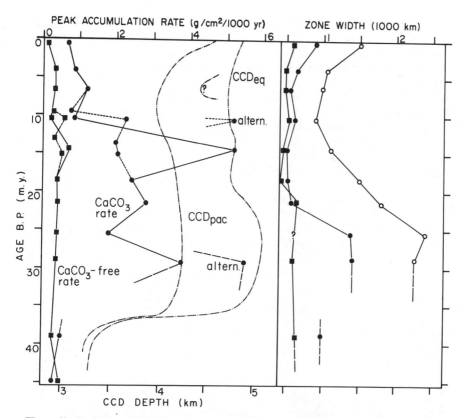

Figure 42. Variation with time of peak accumulation rate and width of equatorial zone of maximum sedimentation during the Cenozoic. Squares: carbonate-free accumulation rate and width of noncalcareous equatorial zone; black dots: carbonate accumulation rate and width of carbonate zone; open circles: width of carbonate zone at zero accumulation rate. Peaks at 10 to 11 and 28 to 30 m.y. B.P. labeled "alternate" indicate high values at sites 70 and 77. Data from Table 4. Dashed curve is based on modified time scale of Figure 35 and data from Table 6. Solid lines connect points based on Berggren (1972a) time scale. Pacific CCD (CCD$_{pac}$) and equatorial CCD (CCD$_{eq}$) added from Figure 29. Modified-time-scale points were calculated for new zones of maximum and minimum sedimentation in Figure 36 using plots of rate versus paleolatitude similar to Figures 40 and 41.

is a function of the dissolution gradient (Heath and Culberson, 1970). This gradient has not remained constant with time (see Chap. 7), and thus the relationship between depth and carbonate accumulation should vary with time as well as with latitude. The small number of curves for this relationship that can be constructed from the available data is inadequate to examine it in detail and to estimate paleodepth corrections that could be applied to data such as those of Table 4. Fortunately, such corrections appear to be less important than might be anticipated, primarily owing to the modest increase of the rate with decreasing depth.

Regional change in depth is not the only factor capable of producing longitudinal gradients in carbonate content and rate of accumulation. The isopleth maps of Figures 15 (sediment isopachs), 25 (carbonate content), 32 (sedimentation rates), and 37 to 39 (accumulation rates) show westward gradients that can only in part be due to a relationship between carbonate content and depth. In Figure 44, longitudinal changes in carbonate content, carbonate accumulation rate, and carbonate-free accumulation rate along the paleoequator are shown in an age-paleolongitude coordinate system. If data from Appendix 3 are used directly, in a manner analogous to Figure 26, a large scatter results from the steep north-south gradients across the equatorial zone. Instead, we have used the paleolongitudes of the westernmost closed ends of all isopleths from the maps of Figures 25, 38, and 39, modified for the alternate middle Miocene time scale of Figure 35. This approach introduces some contouring bias from the base maps but maximizes the use of information from all drill sites.

The carbonate-content diagram (Fig. 44, top) shows a narrow principal $CaCO_3$ gradient far west of the crest of the ancestral East Pacific Rise. Its position parallels the paleodepth isopleths of 4,500 and 5,000 m. The precise relationship between paleodepth and carbonate content varies with time in accord with changes in the CCD. Even the middle Eocene main gradient, which lies above 3,500 m, appears to be controlled by depth, thus lending support to the assumption of a very shallow CCD at that time. Except for a brief and sharp withdrawal toward the rise crest near the Eocene-Oligocene boundary, the carbonate-rich zone has gradually advanced westward from an initial position about 1,000 km west of the rise crest to its

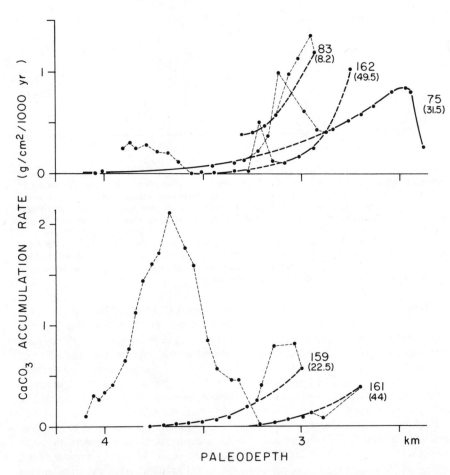

Figure 43. Covariance of carbonate accumulation rate and paleodepth for selected drill sites with a migration history entirely within or entirely outside the equatorial zone. Solid curves fitted by hand to eliminate peaks (dashed lines) caused by temporal variation of carbonate accumulation rate. Numbers in parentheses are basement ages (m.y. B.P.).

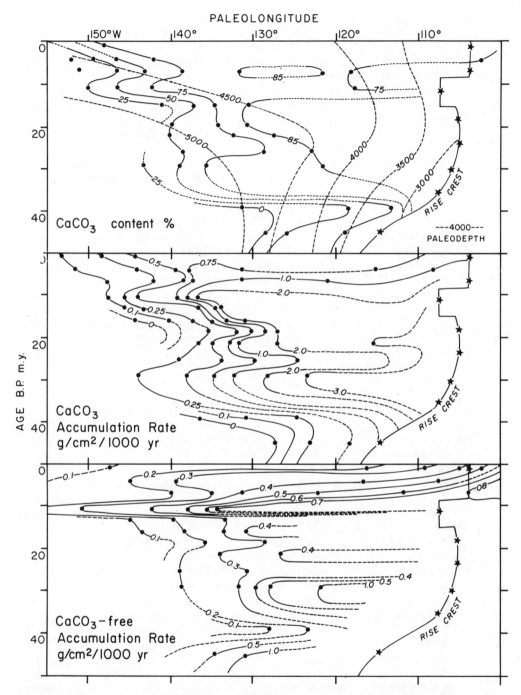

Figure 44. Longitudinal variation with age of carbonate content, carbonate, and carbonate-free accumulation rates along the equator. Based on Berggren (1972a) time scale modified from Figure 35. Data points are longitudes of western closures of contours from Figures 25, 38, and 39, with modifications for altered time scale for 10 to 16 m.y. B.P. Shift with time in position of ancestral East Pacific Rise at lower right interpolated from Figure 21. Paleodepth isopleths (top) based on data points from Figure 21. Diagrams can be visualized as cross sections on the Equator tilted down to the north through all paleoequators.

present position at about 5,000 km. The pattern of depth isopleths shows that this is the result of a gradual broadening of the western slope of the rise during later Cenozoic time (Fig. 23). Thus, the boundary between calcareous and siliceous deposits in the equatorial zone during early Oligocene time was located much closer to the rise crest than at present. Inevitably, the pattern of carbonate accumulation rates (Fig. 44, center) is quite similar and also shows a marked westward migration of the main gradient with decreasing age. However, superimposed on this depth effect is a series of pronounced withdrawals and advances of the main gradient concurrent with the periods of alternating slow and rapid deposition discussed above (Fig. 31).

The main westward gradient of the carbonate-free accumulation rate remains approximately stationary between long. 130° W and 140° W except for a brief but marked eastward withdrawal near the Eocene-Oligocene boundary (Fig. 44, bottom). Several pulses of increased accumulation produced westward advances at the following intervals: 8 to 11, 20 to 22, and around 30 m.y. B.P. The most important of these, at 8 to 11 m.y. B.P., is also reflected in the carbonate accumulation rate. Where both carbonate and carbonate-free pulses occur simultaneously, changes in the supply of biogenic silica and carbonate are implied. Thus, a reduction in the supply or, at least, a supply that extended less far to the west is indicated around 38 to 40 m.y. B.P., and a major increase or extension to the west occurred between 8 and 11 m.y. B.P. The latter may well have also involved an increase in nonbiogenic material such as clay or volcanic ash supplied from the east (Hays, 1972). The large westward advance of carbonate at about 30 m.y. B.P. is not accompanied by a very significant change in the pattern of carbonate-free accumulation rates and should be attributed mostly to a change in carbonate dissolution.

HIGH-FREQUENCY VARIATIONS IN RATES OF DEPOSITION

Temporal variations in the rate of deposition in the central equatorial Pacific thus seem to occur on two separate scales of time. Over the long term, the equatorial rate has changed from very low in early Cenozoic time to high in Oligocene and Miocene time and intermediate to low in the immediate past. Superimposed on this pattern is a more rapid fluctuation from high to low rates on a time scale varying from about 5 m.y. in early Cenozoic time to 1 to 2 m.y. in late Cenozoic time (Fig. 31).

Some drill sites provide evidence of variations of shorter duration. These variations are reflected in rapid changes in the carbonate content on the order of 20 to 30 percent, with associated changes in the weight of calcium carbonate per cubic centimetre. Such changes are displayed only when core recovery and quality are excellent and the sedimentation rate exceeds about 3 m/m.y. Site 83 (Fig. 45), where extremes occur about once every 0.6 m.y., provides the best example. Accumulation rates cannot be determined accurately, because the spacing of peaks is of the same order as the uncertainty limits of the absolute time scale. Similar fluctuations are found in other drill sites (Fig. 46), with spacings ranging from 0.5 to slightly more than 1.0 m.y. (half wave length). The record is incomplete, because stringent conditions must be met to display these variations at all; however, the data show that these fluctuations are not restricted to the equatorial zone and can be observed as far back as middle Eocene time. Similar variations occur in the calcareous section of Late Cretaceous age at site 163. On the other hand, several sites have long sections where the coring record is good enough to show that no such fluctuations exist (Fig. 46).

Lithologic changes of even higher frequency have been described for the equatorial Pacific. Arrhenius (1952) noted large rapid fluctuations of the carbonate content in Quaternary cores, which he explained as the result of alternating periods of cold and warm conditions during the Pleistocene. Hays and others (1969) observed similar fluctuations in the Pliocene record, correlated these with faunal and oxygen-isotope changes in Caribbean cores, and came to a similar conclusion regarding the cause of the fluctuations. The time scale of these fluctuations is shorter than that of the fluctuations described above (Fig. 45). Unit thicknesses ranging from 0.25 to 1.00 m imply tens to a few hundred thousands of years. The amplitudes are on the order of 25 to 50 percent $CaCO_3$.

Similar short-term lithologic fluctuations have been described in pre-Quaternary DSDP cores from the equatorial pacific (Cook, 1972; Hays and others, 1972; van Andel and Heath, 1973b) and were mentioned in Tracey and others (1971).

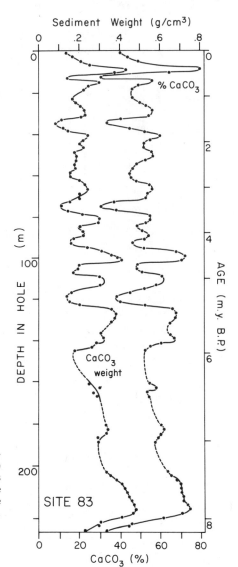

Figure 45. Small-scale variation of carbonate percentage and carbonate weight (g/cm²) with depth in hole and age. Accumulation rates are approximately proportional to carbonate weight but are not indicated because of the uncertainty of interpolating ages within 1-m.y. intervals.

Nannofossil ooze or chalk and radiolarian nannofossil ooze and chalk or, in more extreme cases, nannofossil and radiolarian ooze alternate every 10 to 15 cm. Although the variation is mainly in the carbonate fraction, other sediment components, such as clay and volcanic ash content and abundance of radiolarians, vary as well (Cook, 1972). In the equatorial zone, these fluctuations are usually of late Cenozoic age, but away from the Equator they become progressively older (Cook, 1972), and they have been observed in deposits as old as Eocene (Tracey and others, 1971). Although Hays and others (1972) interpreted them as climatically induced in a manner analogous to their Quaternary equivalents, there is no direct evidence available at the moment that bears on the causes of the variability.

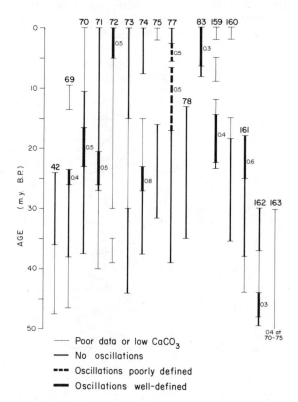

Figure 46. Occurrence of high-frequency fluctuations of carbonate content in all drill sites that have adequate core recovery. Numbers along bars are maximum-to-minimum (half wave length) average durations in millions of years. Thin line indicates poor core recovery, poor core quality, a carbonate content below 25 percent, or an inadequate number of carbonate determinations.

6

Pacific Cenozoic Hiatuses and Erosion

CAUSES AND EFFECTS

The common occurrence of hiatuses in Deep Sea Drilling cores from the Pacific Ocean has frustrated, in part, attempts to recover an unbroken geologic record of the Earth's largest ocean. Their presence is a paradox in that the deep ocean basins are the ultimate sediment catchments, yet a discontinuous stratigraphic record is preserved within them.

Everything that is washed from the continents is carried finally to the ocean bottom; the skeletal material of organisms that live in the sea also must fall to the bottom or be dissolved en route. The fact that hiatuses exist in the record belies the nineteenth-century concept of a still and stagnant abyss. Bottom water has apparently been oxygenated throughout Cenozoic time; thus, a continual renewal of bottom water must have occurred. The presence of hiatuses indicates advective flow of bottom water of sufficient velocity to transport and even erode pelagic sediments.

Recent investigations have demonstrated that the advective flow of bottom water gives rise to well-defined deep currents and that these currents can cause nondeposition, erosion, transportation of sedimentary material, and construction of large depositional features on the sea floor (for example, Amos and others, 1971; Hollister and others, 1974; Johnson, 1972a; Johnson and Johnson, 1970; Jones and others, 1970; Le Pichon and others, 1971; Ruddiman, 1972; Schneider and Heezen, 1966). In the Pacific Ocean, most studies (for example, Burkov, 1969; Knauss, 1962; Panfilova, 1967; Stommel and Arons, 1960; Wüst, 1929) agree that Antarctic Bottom Water moves northward along the western margin of the South Pacific, following the bathymetric contours into the Northern Hemisphere, and then moves northwestward between the Marshall Islands and Mid-Pacific Mountains and eastward through gaps in the Line Island chain (Fig. 47). Thus, the area of this study lies in the eastward flow path of Antarctic Bottom Water. That this arm of the flow causes nondeposition and erosion is evidenced by the widespread exposure of Tertiary sediments on the sea floor (Hays and others, 1969; Riedel and Funnell, 1964). Its effect has been further substantiated by several studies of small areas within this region (Johnson, 1972b; Johnson and Johnson, 1970; Moore, 1970).

Whether erosion, nondeposition, or sediment accumulation occurs within an area is determined by the dynamic balance between the rate at which sediment is supplied to the sea floor and the rate at which it is removed (Fig. 48). In the central equatorial Pacific, where the sediment is predominantly biogenic, the supply rate is largely controlled by the productivity of the near-surface water. The rate of removal is determined by the rate of bottom flow and by the corrosiveness of

73

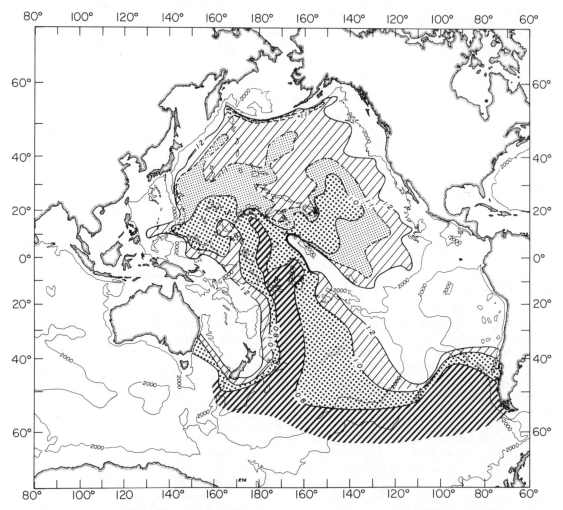

Figure 47. Distribution of bottom potential temperatures (°C) in the Pacific Ocean (after Gordon and Gerard, 1970; Panfilova, 1967).

the bottom water relative to calcite and opaline silica. Continued exposure to bottom water, through nondeposition, might lead to the dissolution of biogenic sediment. This removal of sediment by corrosion is a special case of erosion.

Any change in the productivity of surface water, the rate of bottom-water flow, or the chemical nature of the bottom water can affect the balance between accumulation and erosion and, therefore, the distribution of hiatuses. Tectonic movements associated with plate motions can drastically alter the path of flow or even the source of bottom water and thus change the pattern of occurrence of hiatuses. Finally, in the equatorial Pacific, the northward migration of the Pacific plate can displace a site from an area of accumulation to one of erosion. Therefore, in spite of the paradoxical aspect of hiatuses in the marine record, the problem is not one of finding plausible causes for their existence, but rather finding the probable cause for each occurrence.

The multiplicity of explanations for various cross-sectional patterns of hiatus extent is illustrated in Figure 49. If each hiatus is viewed as resulting from a

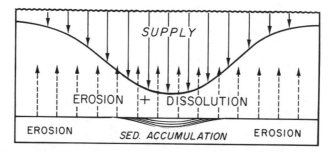

Figure 48. Diagram of the balance of sediment supply and removal rates. Accumulation occurs only where the rate of sediment supply exceeds its rate of removal by erosion and dissolution.

single erosional event, then the time of maximum extent can be defined. The time of cessation can be defined if the overlying material has not been subsequently eroded by a second erosional event. The time of initiation of an erosional event cannot be precisely defined. However, regional synchronism in the age of an unconformity does suggest that the age of initiation is very near the age of the underlying sediment (Berger, 1972).

DEFINITION OF HIATUSES IN DEEP SEA DRILLING SITES

All Pacific Basin sites were used in this compilation. In most cases the existence of a hiatus is clearly defined by the absence of one or more biostratigraphic zones. Only these breaks in the stratigraphic record are accepted as hiatuses and used in the compilation.

In a few cases where the unrepresented period of time is short relative to the average sedimentation rate, it is possible that the section is present but not sampled. Even in continuously cored sections in which the recovered core barrels were full, an average of 35 percent of the section may have been missed (Moore, 1972). This estimate is based on the abnormally frequent occurrence of biostratigraphic boundaries between contiguous cores. Thus, of the 18 m drilled and sampled in two adjacent cores, 6.3 m (0.35 × 18 m) of the actual section could be missed

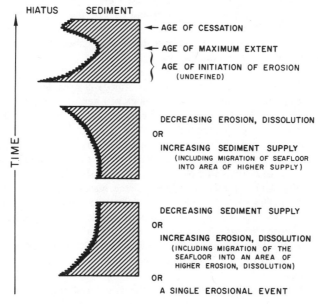

Figure 49. Diagrammatic cross sections of hiatus occurrences. For each erosional episode only the ages of cessation and maximum extent can be accurately defined. Several possible explanations exist for the occurrence of a hiatus in a section of pelagic sediment.

in the drilling and recovery process. If the average sedimentation rate at a site were 2 m/m.y., a hiatus of over 3-m.y. duration could be created by the vagaries of the sampling technique.

In the data set used here, hiatuses of less than 6-m.y. duration were carefully examined. If such a hiatus occurred within a single core it was accepted as real. If it occurred between cores and was of a duration less than 6.3 m divided by the average sedimentation rate (in m/m.y.), then it was considered possible that the gap was created by the drilling technique; the existence of the hiatus was therefore questioned, and it was not included as a reliable hiatus in the compilation.

At sites that were not continuously sampled, hiatuses are considered to exist only when the average sedimentation rate was significantly less than 1 m/m.y.. Where average sedimentation rates were low (1 to 2 m/m.y.) or the preservation of microfossils did not permit accurate stratigraphic definition, hiatuses may be suspected but are not included as such in this compilation.

There are parts of the stratigraphic record in which the present zonation of the various microfossil groups does not provide sufficient resolution to define a hiatus of a few million years. However, the range of certain species may give a clear indication at some sites that a part of the record has been removed or intensively reworked. This is particularly true of upper Eocene intervals, which biostratigraphers in the Pacific have tended to place within a single zone (for example, Dinkelman, 1973; Bukry, 1973). On the basis of the first appearance of such radiolarian species as *Lophochytris jacchia* and *Theocyrtis tuberosa* (Econe variety) and the last appearance of *Podocyrtis goetheana*, a late Eocene hiatus may be suspected in certain drill sites; but again, they are not included as such in this compilation.

TEMPORAL DISTRIBUTION OF HIATUSES

In order to determine the temporal variations in hiatus abundance for deposits of the central equatorial Pacific, only the sites that might be affected by the eastward-flowing tongue of bottom water were considered (Fig. 47). These 40 locations lie south of lat 47° N and east of the Hess Rise and Line Islands (sites 32 to 44, 67 to 84, 155 to 164, and 171 to 173). They include stations that lie outside the tropical region and, therefore, exceed the number of sites for which accumulation-rate data are presented.

For each of 50 1-m.y. intervals (from 0 to 50 m.y. B.P.) the number of sites represented by a hiatus (as here defined) is recorded. This number is expressed as a percentage of the total number of sites at which the particular age interval was sampled. Thus, for any given age, Figure 50 shows the proportion of sites represented by a hiatus.

From 50 m.y. B.P. to the present there are two intervals during which hiatuses are generally abundant: the late Eocene to early Oligocene and the middle Miocene. These broad peaks may be composed of several closely spaced but discrete intervals of abundant hiatuses. The late Eocene hiatus maximum is at 41 to 42 m.y. B.P. and appears to follow closely a distinct minimum in the occurrence of hiatuses at 45 to 47 m.y. B.P. Subsequent to the late Eocene maximum, the occurrence of hiatuses decreased to a second minimum at 35 m.y. B.P. Following this, hiatuses gradually became more common and reached the second major peak in abundance at 11 to 13 m.y. B.P. A final, minor peak occurred near the Pliocene-Quaternary boundary.

A lithologic cross section of the equatorial region (Fig. 51) reveals many of

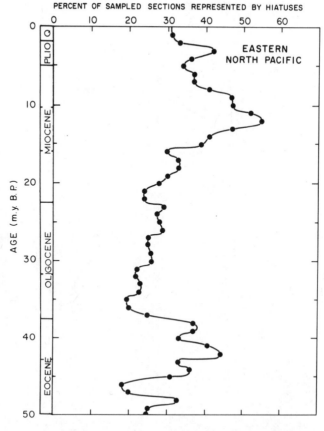

PERCENT OF SAMPLED SECTIONS REPRESENTED BY HIATUSES

EASTERN
NORTH PACIFIC

Figure 50. Occur-
rence of hiatuses in the
eastern North Pacific.
The abundance of hia-
tuses in DSDP sites from
this area is expressed as
the proportion of sam-
pled sections. Data
calculated for each 1-
m.y. increment (Moore
and others, 1975). Time
scale after Berggren
(1972a).

these same features. An extensive hiatus may span the equatorial Pacific in the upper Eocene sediments. A broad area of carbonate-sediment accumulation is noted in lowermost Oligocene deposits. In middle Miocene and late Pliocene time, sediment accumulation was restricted to within 5° of the Equator.

The plot of hiatus occurrences versus time (Fig. 50) has also been compared with a similar diagram for the sites located in the southwestern Pacific—within the region where Antarctic Bottom Water now turns northward into the Pacific Basin (Fig. 52). The distribution of hiatuses relative to time in this region is somewhat similar to that of the equatorial area, particularly for the early part of the record; however, the occurrence of hiatuses is generally more common in the southern region. This may be due to an overrepresentation of sites with shallow paleodepths in the northern region (Berger, 1972). The middle Eocene hiatus minimum is again followed closely by a late Eocene to early Oligocene peak. Although the onset of abundant hiatuses is identical in the two areas, the abundance peak is much broader in the southern region, and the maximum is reached somewhat later (39 to 40 m.y. B.P.). It has been suggested that this peak in hiatus abundance is associated with a major global cooling and marks the beginning of the extensive production of sea ice and the formation of cold Antarctic Bottom Water (Kennett and others, 1975).

The early Oligocene minimum in hiatuses found in the central equatorial Pacific is missing in the southern area. Instead, the occurrence of hiatuses decreases from the late Eocene until two successive peaks in abundance are encountered in the

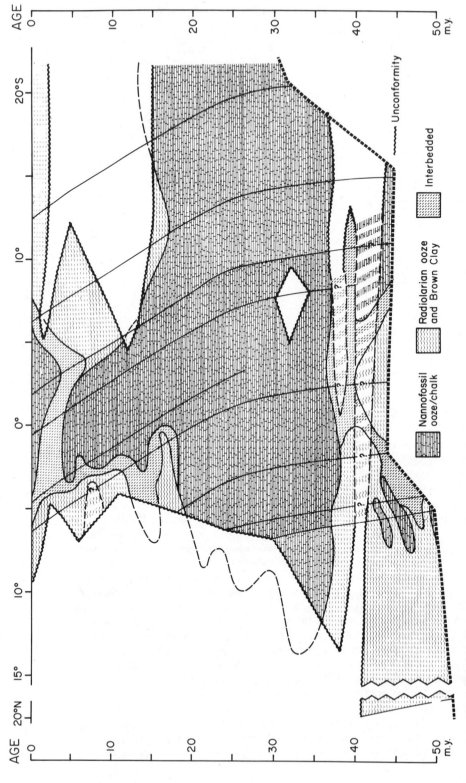

Figure 51. Lithologic cross section of the central equatorial Pacific. Lithology plotted against time (Berggren, 1972a) in a longitudinal cross section based on nine sites migrated through time (see Chap. 3). Diagonal lines indicate location of sites 161, 162, 42, 70, 71, 72, 73, 74, and 75. Unconformities bounded by wavy lines. Dashed lines, question marks, and broken patterns indicate possible unconformities.

Neogene: one at about 18 m.y. B.P. and the other at 10 to 11 m.y. B.P. The former is matched only by a small oscillation in the curve for the northern area, whereas the latter corresponds to the second major peak in hiatus abundance shown in Figure 50. The minor Pliocene-Quaternary peak seen in the northern area appears only as a shoulder on the broad late Neogene peak in Figure 52.

Although the peaks in hiatus abundance tend to coincide in the two areas, they do not show the same relative importance. The late Eocene episode is by far the most important in the southern area, whereas in the northern region it appears to be of approximately equal importance to the middle Miocene episode. This may be a real difference or may result from either the previously discussed insensitivity of the upper Eocene biostratigraphic sequences in the equatorial Pacific or the different depth distributions of sites in the two areas.

The greater abundance of hiatuses in the southern region causes the peaks in abundance to appear broader than those of the northern area; however, the mean hiatus durations of sites in the two areas are not significantly different, as is shown by the following: southwest Pacific, mean hiatus duration, 14.0, standard deviation, 13.2; North Pacific, mean hiatus duration, 13.0, standard deviation, 14.0.

SPATIAL DISTRIBUTION OF HIATUSES

The periods of maximum abundance in hiatuses are shown in maps of the Pacific Basin representing the intervals 0 to 5 m.y. B.P. (Fig. 53), 10 to 15 m.y. B.P. (Fig. 54), 15 to 20 m.y. B.P. (Fig. 55), and 40 to 45 m.y. B.P. (Fig. 57). The distribution of hiatuses during the early Oligocene and Eocene minima is shown in Figure 56 (30 to 35 m.y. B.P.) and Figure 58 (45 to 50 m.y. B.P.). These maps are based on all Pacific sites containing reliable hiatuses (the same data set from which the compilations on the temporal distribution of hiatuses were made). Those sites at which suspected or questionable hiatuses are found are also indicated on the maps.

Interval from 0 to 5 m.y. B.P.

Three regions in which the entire interval from 0 to 5 m.y. B.P. is represented by a hiatus can be recognized. They are located south of New Zealand and in the eastern and western North Pacific (Fig. 53). They correspond to regions where the bottom water now enters the southern Pacific and turns northward to the Mid-Pacific Mountain–Marshall Island gap and eastward through the Line Island gap (Fig. 53). The exact boundaries of these regions are not well defined in the north; however, in the eastern North Pacific, the southern boundary appears to coincide with the Clipperton Fracture Zone (Johnson, 1972a).

Away from the centers of maximum erosion, the age of cessation of the erosional episode increases, suggesting either a gradual decrease in the intensity of the bottom currents or an increasing rate of sediment supply. In the equatorial area, several sites (65, 70, 74, 158, 166) exhibit a short, distinct hiatus at approximately 2 to 3 m.y. B.P. In the south, only sites 280 (1 to 4 m.y. B.P.) and 281 (1 to 5 m.y. B.P.) have hiatuses that appear to correspond to this event.

Interval from 10 to 15 m.y. B.P.

Eroded areas in the North Pacific reached their maximum extent for Neogene time during the interval from 10 to 15 m.y. B.P. (Fig. 54). A well-defined area

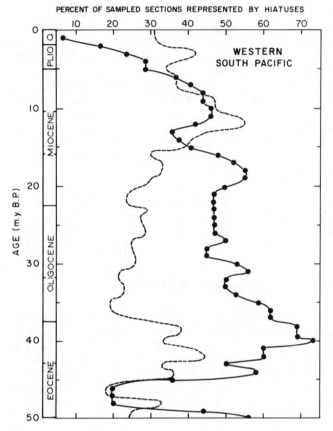

Figure 52. Occurrence of hiatuses in the southwestern Pacific. The abundance of hiatuses in DSDP sites from this area is expressed as the proportion of sampled sections. Data calculated for each 1-m.y. increment (Moore and others, 1975). Time scale after Berggren (1972a).

of erosion also existed in the far southwest Pacific. In all regions, the age of maximum extent of the hiatus appears to be 12 m.y. B.P. South of the Equator, sites 73 and 80 show short, discrete hiatuses at 10 to 13 m.y. B.P., again indicating a widespread erosional event that reached a maximum about 12 m.y. ago. The timing of this event coincides with the earliest recorded ice-rafted debris in site 278 south of New Zealand; the event is probably associated with the development of a large ice cap on Antarctica (Kennett and others, 1975).

Interval from 15 to 20 m.y. B.P.

In the northern regions the eroded areas in the interval from 15 to 20 m.y. B.P. have the same general pattern as during the 10 to 15-m.y. B.P. interval but are of diminished extent (Fig. 55). The narrowing of the equatorial carbonate belt about 16 to 17 m.y. ago (Fig. 51) and the occurrence of discrete hiatuses at two sites (site 65, 14 to 16 m.y. B.P.; site 69, 16 to 20 m.y. B.P.) from the southern part of the tropical region are the only signs of similarity with the pattern of hiatuses in the southwestern Pacific. In the latter area, there was an erosional episode with an apparent age of cessation of approximately 16 m.y. B.P. If this event extended into the northern areas, its effects have been largely obliterated by the event 12 m.y. ago. The absence of hiatuses south of the Equator in the eastern tropical Pacific suggests that the event prior to 16 m.y. B.P. reached its

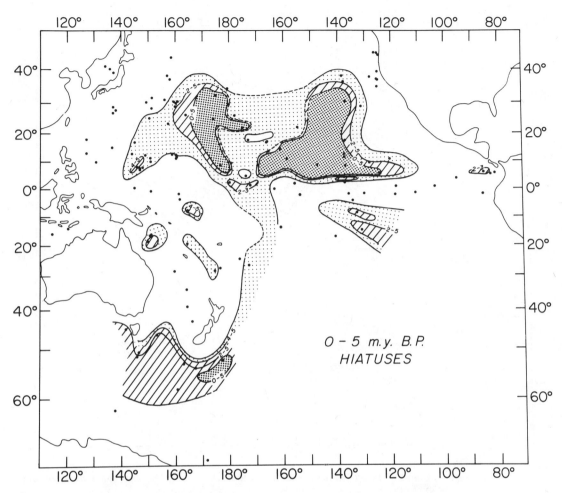

Figure 53. Hiatuses in the Pacific Basin, 0 to 5 m.y. B.P. Large dots indicate site locations. Heavy dotted pattern indicates area in which entire interval is missing. Diagonal pattern indicates sites in which part of the interval is missing. Light dotted pattern indicates sites in which the occurrence of a hiatus is uncertain.

greatest extent in the Tasman and Coral Seas and affected only the southernmost sites of the northern area.

This episode closely follows the initiation of the Antarctic Convergence and the beginning of silica-rich deposition near Antarctica (Kennett and others, 1975).

Interval from 30 to 35 m.y. B.P.

The interval from 30 to 35 m.y. B.P. represents a period with few hiatuses in the eastern North Pacific (Fig. 50). The small eroded area north of the Equator conceivably could be the last remnant of the 12-m.y. B.P. erosional event. South of the Equator, however, there is a short hiatus at site 72 (30 to 34 m.y. B.P.) that may be associated with the widespread hiatus of the southwestern Pacific (Fig. 56). This erosional pattern is thought to be a continuation of the event resulting from the opening of the Australian-Antarctic seaway and associated with the formation of Antarctic Bottom Water (Kennett and others, 1975).

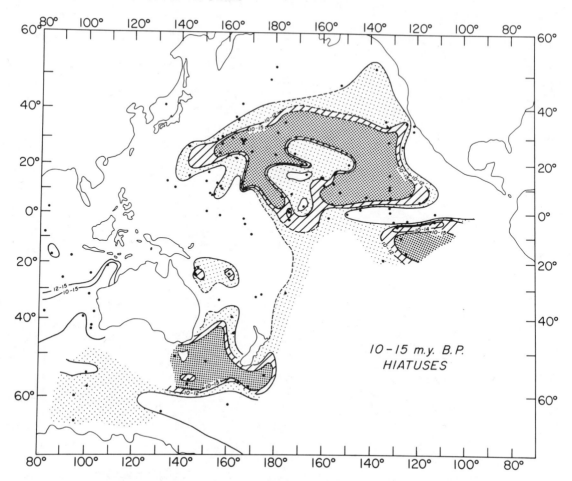

Figure 54. Hiatuses in the Pacific Basin, 10 to 15 m.y. B.P. Symbols and patterns are same as in Figure 53.

Interval from 40 to 45 m.y. B.P.

The interval from 40 to 45 m.y. B.P. has the maximum abundance of hiatuses. In all regions the age of maximum extent is 40 to 42 m.y. B.P. According to Kennett and others (1975) this interval followed the opening of the Tasman seaway to surface-water flow and just preceded major changes in the bottom fauna associated with a sharp cooling of bottom water at the Eocene-Oligocene boundary. The pattern of hiatus occurrence suggests that bottom water may have entered the Pacific Basin north of the Australian block and flowed northward into the North Pacific and southward into the Coral Sea and South Pacific (Fig. 57).

Interval from 45 to 50 m.y. B.P.

The interval from 45 to 50 m.y. B.P. spans the middle Eocene minimum in hiatus abundance and shows the same general pattern of occurrence as the succeeding interval (Fig. 58). Site 164 in the equatorial region probably shows the effect of

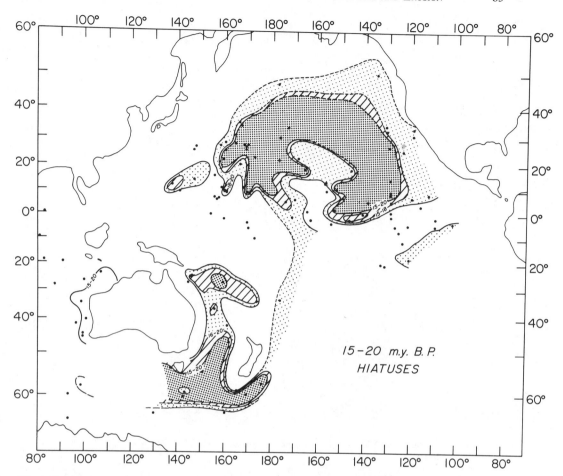

Figure 55. Hiatuses in the Pacific Basin, 15 to 20 m.y. B.P. Symbols and patterns are same as in Figure 53.

the erosional episode at 42 m.y. B.P. All other sites, however, show the effects of an earlier episode that ceased at 47 m.y. B.P.

SUMMARY AND CONCLUSIONS

Within the past 50 m.y. the earliest major erosional episode reached its maximum extent at about 42 m.y. B.P. in the central equatorial Pacific and at about 40 m.y. B.P. in the southwestern Pacific. Surprisingly, the age of the underlying sediments is 44 to 45 m.y. in all areas. The areal distribution of the hiatuses suggests that bottom water flowed into the Pacific from the tropical region of the Indian Ocean and spread from there south into the Coral Sea and South Pacific and northward into the western North Pacific. It is not clear whether a single clockwise gyre was formed in the North Pacific or two separate gyres formed in the eastern and western basins.

Comparison with accumulation-rate data for the equatorial Pacific (Fig. 42, App. 3)

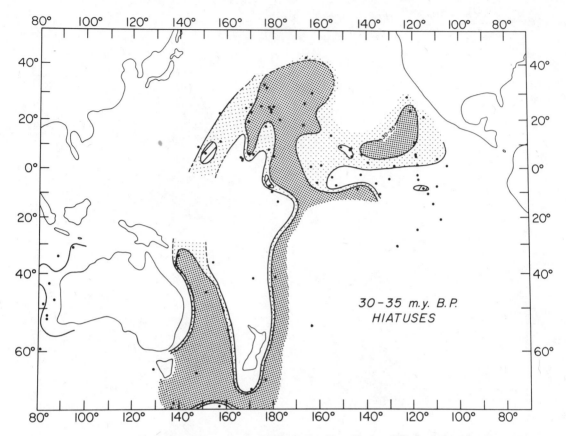

Figure 56. Hiatuses in the Pacific Basin, 30 to 35 m.y. B.P. Symbols and patterns are same as in Figure 53.

indicates that late Eocene time was a period of moderately rapid carbonate-free sediment accumulation but relatively slow carbonate accumulation. Such a pattern suggests fairly high productivity, with bottom water that is relatively old (that is, corrosive with regard to calcite).

The timing of the late Eocene maximum in hiatus abundance is close to that of the separation of the South Tasman Rise and Antarctica and may coincide with the opening of the Drake Passage. Evidence bearing on the exact time of opening of the Tasman passage to deep flow is sparse and subject to different interpretations. Paleomagnetic data from this region (Weissel and Hayes, 1972) do not provide a definitive answer. Kennett and others (1972, 1975) suggested that total separation of Antarctica and the South Tasman Rise did not occur until late Oligocene time. The sea floor directly south of the South Tasman Rise, however, is at least late Eocene in age, and a deep connection between the South Australian and Tasman Basins could have existed through the fracture zones that join the spreading centers located east and west of the passage. A similar situation exists today between the eastern and western Atlantic basins, which have their deepest connection through the Romanche Fracture Zone.

With continued sea-floor spreading, the width and depth of this connection would have increased. Thus, the DSDP sites located to the north and west of this highly channelized flow (sites 280, 282) might not have been affected by eroding currents

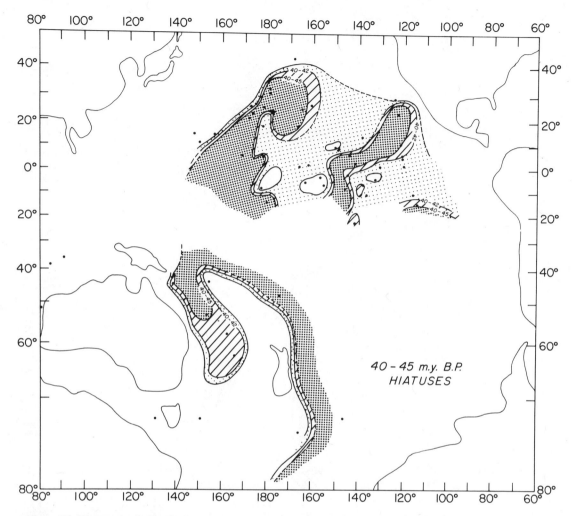

Figure 57. Hiatuses in the Pacific Basin, 40 to 45 m.y. B.P. Symbols and patterns are same as in Figure 53.

until the sill depth of the Tasman passage reached the depth of the sea floor at each site location.

Practically no data exist on the exact timing of the opening of the Drake Passage. It is clear, however, that the Eocene-Oligocene boundary marks the time of a major global cooling, a marked cooling of both surface and deep waters in high southern latitudes, the production of sea ice, a pronounced change in benthic and planktonic faunas, and the production of bottom water that was less corrosive with regard to calcite (Kennett and others, 1975; van Andel and Moore, 1974). It is suggested that these major oceanographic changes resulted from the initiation of the circum-Antarctic current.

Kennett and others (1975) argued that global cooling led to the development of glaciers on Antarctica, which in turn induced the formation of cold bottom water in the area of the Ross Sea. It seems likely to us that the development of the circumpolar current gave rise to this marked global cooling through the increased residence time of surface water in high southern latitudes (Moore and

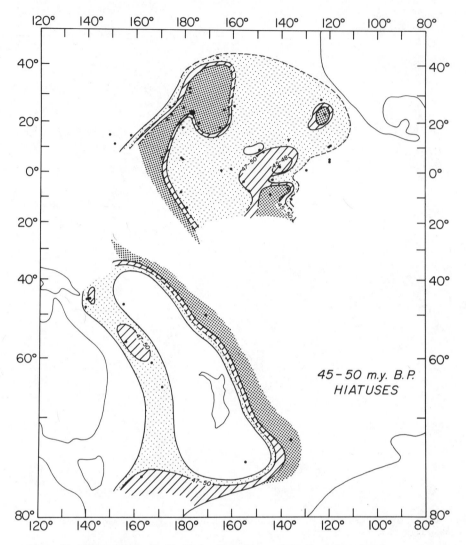

Figure 58. Hiatuses in the Pacific Basin, 45 to 50 m.y. B.P. Symbols and patterns are same as in Figure 53.

others, 1975) and that the growth of this current led directly to both an increase in the glaciation of Antarctica and the rapid formation of cold bottom water in high southern latitudes. The pronounced change in the chemical nature of the bottom water at the Eocene-Oligocene boundary would thus be a result of a shift in the locus of bottom-water formation from the North Atlantic–Arctic region to Antarctica (Moore and others, 1975).

The details of the shift are poorly understood; however, the maximum in hiatus abundance occurred at about 42 m.y. B.P. in the eastern North Pacific (Fig. 50) as well as in most other major ocean basins (Moore and others, 1975). In the western South Pacific and eastern Indian Ocean (Fig. 52; Moore and others, 1975) the maximum of the hiatuses is much nearer the Eocene-Oligocene boundary (37 to 40 m.y. B.P.). This difference, together with the change in the spatial distribution of hiatuses over the Eocene-Oligocene boundary (see Figs. 56, 57), suggests a shift in the path of bottom-water flow into the Pacific from north of the Australian block in Eocene time to a route south of (or just east of) Australia in Oligocene time. When the central equatorial Pacific had a minimum number of hiatuses and a rapidly increasing rate of carbonate and noncarbonate accumulation (30 to 35 m.y. B.P.), the entire western part of the Pacific was undergoing extensive erosion. The initial path of northward flow may have passed through the Tasman and Coral Sea Basins; however, later in the Tertiary the flow apparently passed only east of New Zealand (Figs. 53 to 56).

The 18-m.y. B.P. peak in hiatus abundance in the southwestern Pacific (Fig. 52) has only a small counterpart in the equatorial region; it does coincide with a declining rate of carbonate accumulation (Fig. 42). This peak also depicts a time of increased deep-water flow from the North Atlantic and a time of peaks in hiatus abundance in other ocean basins (Moore and others, 1975). In both the southwestern and equatorial Pacific, oxygen-isotope measurements of planktonic and benthic foraminifers (Douglas and Savin, 1971; Shackleton and Kennett, 1975a) suggest a slight warming at this time, particularly of surface water. If this warming led to a greater vertical stability of the water column, a less rapid overturn of bottom water and the development of bottom water that was more corrosive with regard to calcite may have occurred.

The major Neogene erosional event for the equatorial Pacific reached its maximum extent in the North and South Pacific about 12 m.y. ago. It correlates with a major build-up of the eastern antarctic ice cap, which is signaled by abundant ice-rafted debris in the sediment south of New Zealand (Kennett and others, 1975).

For most of middle and late Tertiary time, the abundance of hiatuses tends to be negatively related to the carbonate accumulation rate on the Equator (see Figs. 50, 52, 42); with increasing rates of accumulation, the abundance of hiatuses generally decreases. The 12-m.y. B.P. episode appears to be anomalous in that the high abundance of hiatuses is matched by high equatorial accumulation rates. This distinctive event apparently involved a high rate of bottom-water formation (bottom water that was not particularly corrosive to calcite) and high productivity in the equatorial region. It has been suggested that it marks the time when true Antarctic Intermediate and Bottom Waters were first formed and the present vertical structure of the oceans was first established (Moore and others, 1975).

The last peak in hiatus abundance in the equatorial Pacific occurred at 2 to 3 m.y. B.P., and it is not clearly represented in the southwestern Pacific. It approximately coincides with the beginning of the Pleistocene and the beginning of ice-sheet development in the Northern Hemisphere (Berggren, 1972b; Shackleton and Kennett, 1975b).

There were clearly two major episodes of erosion in the equatorial Pacific during the past 50 m.y. One was associated with the development of a circum-Antarctic current and a change in the flow path of bottom water, and the other with the development of the antarctic ice sheet and the formation of the Antarctic Bottom Water. The linking of these events with the sedimentation in the equatorial Pacific indicates the need to study both deep and surface circulation of the whole ocean when dealing with the detailed history of a particular region.

7

Paleoceanography of the Central
Equatorial Pacific Ocean

The information contained in the preceding chapters can be summarized from several interrelated but distinct points of view. In this chapter, we review the data with regard to the depositional history of the central equatorial Pacific and then evaluate information on the calcite compensation depth and carbonate accumulation rates in terms of the evolution of supply and dissolution of calcium carbonate during Cenozoic time. Finally, we attempt to place the history of the equatorial Pacific in the context of available information on the global evolution of oceanic surface and deep-circulation patterns.

INTRODUCTION AND REVIEW

In Chapter 2 we showed that two processes dominate the deposition of sediment in the equatorial Pacific: the production of biogenic calcareous and siliceous material, which is maximal at the Equator, and the dissolution of carbonate material with depth. As a result, the type of sediment accumulating at any point on the sea floor at any given moment is a function of its position relative to the Equator and of its depth. Biological productivity, carbonate dissolution as a function of depth, and sea-floor depth all vary with time. Another factor, the input of nonbiogenic material such as volcanic ash and terrigenous clay, cannot be fully evaluated with present data but seems on the whole to play a minor role.

It is useful to emphasize once more that the data and conclusions depend critically on several basic assumptions, the most important of which is the validity of the absolute chronology used. The use of the Berggren (1972a) time scale (in this chapter modified with the alternate scale of Fig. 35) has produced consistent results, such as the parallelism between the CCD and sedimentation-rate curves. This internal consistency and the surprisingly small random variation are the best arguments we can advance at this time for our confidence in the absolute chronology. A second critical parameter is plate rotation. In Chapter 4 we noted some significant differences in the history of the calcite compensation depth as given in this study and the one described by van Andel and Moore (1974). The differences, which are not trivial, result mainly from the use of different schemes for the reconstruction of the drill-site migration tracks. This illustrates vividly the critical role of the underlying assumptions about plate tectonics. The same is true, although to a lesser degree, of the reconstruction of paleodepth.

The evolution of the calcite compensation depth and the temporal variations of the peak accumulation rate and width of the equatorial zone of maximum deposition are closely correlated (Fig. 42). The most striking feature is the large drop of the CCD at the Eocene-Oligocene boundary, which is followed by a large increase in the width and accumulation rate of the equatorial carbonate zone. Until 20 m.y. B.P., a wide equatorial zone of rapid deposition persisted, followed by a gradual decline paralleled by a large rise of the equatorial CCD. The CCD for the Pacific did not rise until somewhat later. During the past 5 to 7 m.y. the Pacific CCD has again sunk to greater depth, followed closely by a decrease in the width of the equatorial zone, but it has not been matched by an increase in the carbonate accumulation rate.

The large change in the position of the CCD at the Eocene-Oligocene boundary was probably worldwide and resulted in a change from dominantly siliceous sediment in Eocene time to dominantly calcareous sediment in Oligocene time. This conspicuous transition, observed in many parts of the world, has sometimes been interpreted in terms of uplift and subsidence of the sea floor (Frerichs, 1970) rather than as evidence for changes in the position of the depth of carbonate solution.

The variation with time of the CCD, which is responsible for major lithologic changes in equatorial sediment, is the result of temporal variations of the following more fundamental factors: the supply of carbonate from surface water, the depth of the lysocline, and the gradient of the carbonate dissolution rate. These, in turn, are indicators of oceanographic and climatic changes, such as variations in the intensity of upwelling in the equatorial convergence, evolutionary and environmental changes affecting planktonic assemblages, and temporal variations in the nature and circulation of intermediate and deep waters. Most of our data bear mainly on the dissolution of carbonate at depth and, hence, on the nature of intermediate- and deep-water circulation rather than on surface fertility. Information regarding the deep circulation is also recorded in the unconformities and hiatuses discussed in Chapter 6. However, limited information regarding surface production and, hence, some insight into surface-circulation changes can be derived from the depth dependence and longitudinal variation of carbonate accumulation rates and from the carbonate-free accumulation rates. The lack of knowledge regarding the composition of the latter, however, limits their usefulness in this regard.

The broad pattern of change during the past 50 m.y. is perturbed by fluctuations of the carbonate content and deposition rates on several shorter time scales (Figs. 31, 45). These fluctuations, although often so described, do not appear to be truly cyclic. Arrhenius (1952) attributed the very short term Quaternary variations to climate-induced changes in productivity, but Berger (1973c) ascribed them to short-term variations in solution. If either of these views is correct, the occurrence of similar variations as early as Eocene time implies major climatic oscillations on several time scales. J. Imbrie (1974, oral commun.) argued that such complex climatic periodicities may result from the interaction of several periodic variations in the Earth's orbital parameters.

CARBONATE SUPPLY AND DISSOLUTION

Ever since Arrhenius (1952) reported large variations in the abundance of carbonate in Quaternary sediments of the equatorial Pacific, the relative importance of fluctuations of carbonate supply (production) and dissolution in generating such variations has been debated. Arrhenius conceded that his cores showed evidence of dissolution, but he ascribed the variations to changes in surface productivity. He suggested that periods of glaciation were characterized by intensified equatorial

circulation, which produced greater upwelling and led to carbonate enrichment
of glacial-age deposits on the sea floor. Pisias's (1974) study of opal accumulation
in the easternmost equatorial Pacific provides some support for the idea of increased
glacial productivity. Recently, however, Berger (1973c) re-examined Arrhenius's
data and concluded that variations in the ratios of planktonic to benthic foraminifers,
of planktonic foraminifers to radiolarians, and of resistant to easily dissolved
planktonic foraminifers are all much more indicative of changes in dissolution
than of changes in production. In fact, he believed that virtually all of the variation
in carbonate content in Arrhenius's cores reflects changes in the degree of dissolution,
glacial deposits being less dissolved than interglacial ones.

The arguments that have developed over the Quaternary cores are equally
applicable to the Cenozoic sections considered in this study. Because we lack
quantitative faunal and floral data, however, we cannot employ Berger's (1973c)
methodology to distinguish productivity (or supply) variations from changes in
dissolution. From van Andel and Moore (1974) and Chapter 4, however, it is clear
that either or both of these factors have changed dramatically during the past
50 m.y.

Our attempt to distinguish variations in supply from those of dissolution depends
on their differing influences on the rate of accumulation of calcium carbonate
as a function of paleodepth. The relationships are obvious from a simple model
of dissolution in the equatorial Pacific (Fig. 59). Peterson's (1966) experimental
data suggest that dissolution is minimal above a depth close to Berger's lysocline
(Parker and Berger, 1971). Below this level, the rate of dissolution appears to
increase more or less linearly with increasing depth (Peterson, 1966; Heath and
Culberson, 1970), as shown by curve D of Figure 59a. If the rate of supply of
calcite from shallow water at time A is S, then the rate of accumulation of carbonate
as a function of water depth will be given by curve A(1), and the CCD(A) will
be the calcite compensation depth at that time. If, at time B, the CCD has moved

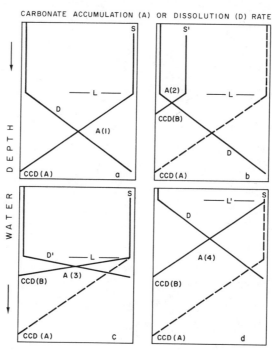

CARBONATE ACCUMULATION (A) OR DISSOLUTION (D) RATE

**Figure 59. Simplified model
of carbonate dissolution (D) and
rate of accumulation (A) in the
equatorial Pacific. CCD = calcite compensation depth; L =
lysocline; S = rate of supply
of calcite. A shift of the CCD
from level CCD (A) to CCD (B)
results from a decrease in S to
S' (b), an increase in the depth
gradient of the dissolution rate
from D to D' (c), or a rise in
the lysocline from L to L' (d).**

to depth B, changes in any of the three parameters—rate of supply of calcite, rate of dissolution as a function of depth, or depth of the lysocline—could be responsible. The three extreme situations are shown in Figures 59b through 59d. In Figure 59b the change in CCD is entirely due to a change in supply, in Figure 59c to a change in the dissolution gradient, and in Figure 59d to a change in the depth of the lysocline. The number of permutations possible is infinite if all three parameters are allowed to vary.

It is clear that knowledge of the variation of the rate of carbonate accumulation as a function of depth is sufficient to define all three parameters. Figure 60 shows the depth dependence of the carbonate accumulation rate at 18 to 20 m.y. B.P. (during early Miocene time) and at 35 m.y. B.P. (during early Oligocene time) shortly after the CCD dropped abruptly from its shallow Eocene depth (Chap. 4; van Andel and Moore, 1974). In each case, samples less than 3° and 4° to 10° of latitude from the paleoequator have been segregated. Samples 3° to 4° from the paleoequator and east of paleolongitudes 115° W (at 18 to 20 m.y. B.P.) and 110° W (at 35 m.y. B.P.) have been omitted to minimize scatter due to longitudinal and latitudinal supply gradients (Figs. 40, 44).

The plots of accumulation rate versus depth are approximately linear in both cases (and, in fact, for all similar plots we have studied), which supports the simplified model of Figure 59. The plots can be described by equations of the following type: accumulation rate = $S - D \times$ water depth, where S is the "extrapolated surface production" (the rate of supply at the surface if the lysocline did not exist), and D is the vertical gradient of the rate of carbonate dissolution (mass / unit area / unit time / unit depth increase). Given the possible complications due to lateral sediment movement caused by topographically controlled bottom currents at each site (Johnson and Johnson, 1970; Moore, 1970), the scatter of points in Figure 60 is surprisingly small.

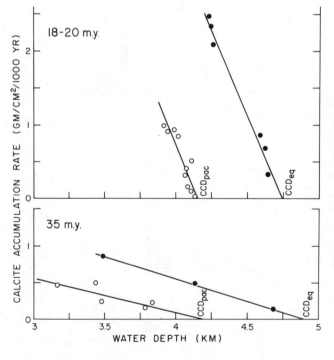

Figure 60. Variation in the accumulation rate of calcite as a function of paleodepth for samples within 3° of the Equator (solid circles) and 4° to 10° from the Equator (open circles) during early Oligocene time (35 m.y. B.P.) and early Miocene time (18 to 20 m.y. B.P.). The time variation of the slope of the equatorial (lower) best-fit lines is shown in Figure 61b.

The rate of dissolution during early Miocene time (4.6 to 4.8 g/cm^2/1,000 yr/km) was almost an order of magnitude greater than the rate during early Oligocene time (0.45 to 0.6 g/cm^2/1,000 yr/km). Unfortunately, the depth distribution of samples is too limited to define the lysocline (knickpoint) in the dissolution curve (Fig. 59a), so that Figure 60 cannot be used to estimate the rate of input of calcite from surface water. Similarly, data like those of Figure 60 cannot show whether a change in the CCD resulted from a productivity change (Fig. 59b) or a shift in the level of the lysocline (Fig. 59d). Because the rise crest at about 3,000 m represents the upper limit of the sample depths for the area we have studied, it is doubtful whether good estimates of calcite supply during the Cenozoic can be recovered from the DSDP data in the eastern Pacific. Knowledge of the carbonate content of the incoming sediment would allow the initial supply rate and lysocline to be calculated from curves like those of Figure 60, but such information is not available and cannot be estimated. For example, in the lat 3° S to lat 3° N region for the period 18 to 20 m.y. B.P., the highest carbonate value is 83 percent for calcite that accumulated at a rate of 2.47 g/cm^2/1,000 yr. If the incoming sediment contained 95 percent carbonate, then from Berger's (1971) expression the percentage that must have dissolved (L) to yield our sample is

$$L = 100\left(1 - \frac{R_0}{R}\right) = 100\left(1 - \frac{5}{17}\right) = 70.6\%,$$

where R_0 is the initial noncarbonate content, and R is the noncarbonate content of our sample.

For this R_0, the rate of supply was 8.4 g/cm^2/1,000 yr (100/29.4 × 2.47), and the lysocline depth was 2,900 m—both plausible values. If, however, the incoming sediment contained 98 percent carbonate ($R_0 = 2$), then L = 88.2 percent, the supply rate was 21 g/cm^2/1,000 yr, and the lysocline was at 170 m. Clearly, we have no way of choosing the correct value of R_0. Only if the depth distribution of our samples had spanned the lysocline could we pin down changes in either or both the rate of supply and lysocline depth during the Cenozoic. Perhaps in the western Pacific, where a number of sites have been drilled on shallow rises (Magellan, Manihiki, Ontong-Java), the carbonate dissolution patterns of the past will be fully recoverable.

Even though direct evidence on changes in the lysocline depth and rate of carbonate supply is lacking, the constancy with which noncarbonate detritus has accumulated in the central Pacific during the past 50 m.y. (Fig. 42) suggests that the supply rate has varied by no more than a factor of two or three. M. Leinen (1975, written commun.) is determining the biogenous (opal) and nonbiogenous contributions to the noncarbonate material, but as yet we can only make the conservative assumption that opal is the dominant noncarbonate component of the equatorial samples. Consequently, variations in the rate of accumulation of noncarbonate materials should be a good indicator of variations in the rate of input of biogenous sediment (Pisias, 1974).

Figure 61 is an attempt to summarize the relationships between carbonate accumulation rates within 3° of latitude of the paleoequator and paleodepths during the past 45 m.y. The dissolution and extrapolated surface-production curves are the D and S values as a function of geologic age of the following equation: accumulation rate = S − D × water depth. The equatorial CCD curve is the depth at which the accumulation rate from the equation extrapolates to zero. Because our earlier CCD curves were defined as 20 percent carbonate, the curves of Figures 28 and 61 differ in detail. Figure 61 should show the best estimate, because more information is utilized for each age interval than in earlier figures. The depths of the accumulation-rate isopleths at 2 and 6 g/cm^2/1,000 yr (Fig. 61) are calculated from the above equation. The present equatorial lysocline appears to be close to 6 g/cm^2/1,000 yr, whereas 2 g/cm^2/1,000 yr is a typical value for the carbonate accumulation rate in modern equatorial sediments.

It is clear from Figure 61 that the impression of stable carbonate deposition during the past 35 m.y. suggested by the relatively uniform equatorial CCD is misleading. The gradient of the rate of dissolution with depth has varied by almost a factor of 30 during this period—a variation that has been largely counteracted by changes in supply rate and lysocline depth.

The sequence of events suggested by Figure 61 is as follows:

1. The rapid drop in the equatorial CCD at the end of the Eocene was due to a roughly threefold drop in the vertical gradient of the dissolution rate. The rate of supply apparently decreased (Fig. 61c), but not enough to maintain the CCD at its Eocene level. Alternatively, the lysocline could have been close to the depth of the rise crest in late Eocene time and have risen almost to the surface during early Oligocene time.

2. During the Oligocene, the supply rate generally increased (or the lysocline dropped) to a maximum at about 30 m.y. B.P. This maximum is also seen in the noncarbonate accumulation rate (Fig. 42), but its effect on the equatorial CCD was largely offset by an increased vertical gradient in the dissolution rate. A slight decrease in supply (or rise of the lysocline) occurred at about 26 m.y. B.P. during late Oligocene time.

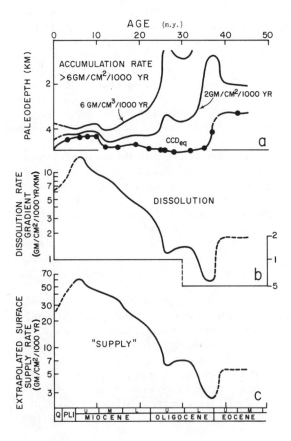

Figure 61. Temporal variation of the CCD computed from the relationship of accumulation rate to depth (Fig. 60), of the vertical gradient of the dissolution rate, and of the "extrapolated surface carbonate supply" during the past 50 m.y. The temporal variation of the CCD is obtained by extrapolating the accumulation-rate curves of Figure 60 and similar ones for other indicated intervals to zero. The accumulation-rate isopleths for 2 and 6 g/cm^2/1,000 yr (top) are calculated similarly. Center (b) and bottom (c) figures are time variations of the coefficients D and S of the following equation: accumulation rate = S − D × paleodepth, where D is the vertical gradient of the dissolution rate below the lysocline, and S is the "extrapolated surface supply"—the rate of influx of carbonate if there were no lysocline.

3. During latest Oligocene and early Miocene time (26 to 16 m.y. B.P.) the supply rate increased, the lysocline dropped to close to its present level, and the vertical gradient in the dissolution rate increased about fivefold (Fig. 61). However, the equatorial CCD rose only 200 to 300 m, because the increased dissolution was largely cancelled by the increased supply or by the depression of the lysocline.

4. From middle Miocene time to the present, the supply rate has been fairly uniformly high. During late Miocene time, the vertical gradient of the dissolution rate increased enough to raise the equatorial CCD to about 4,350 m from 10 to 6 m.y. ago.

5. The present regime is comparable to the pattern for the past 15 m.y., although, surprisingly, knowledge of absolute accumulation rates for late Quaternary time lags behind knowledge for much of the Tertiary.

The question arises as to the reliability of the patterns deduced from Figure 61. The slope of the dissolution curve is particularly sensitive (in fact, directly proportional) to errors in the duration of 1-m.y. intervals. Thus, the previously discussed cautions about uncertainties in the age assignments used in this synthesis bear repeating. The lack of rapid temporal changes in the gradient of the dissolution rate, however, suggests that the intervals used are internally consistent.

The oceanographic implications of Figure 61 must be speculative inasmuch as the functional relationships of carbonate chemistry, water structure, and deep circulation to calcite dissolution in today's ocean are still not understood. It is striking, however, that during the past 35 m.y. the equatorial CCD has fluctuated only a few hundred metres, whereas the vertical gradient of the dissolution rate has varied by almost a factor of 30. The complementary changes in the dissolution gradient, the supply rate, and the position of the lysocline point to the existence of a strong negative feedback, which tends to hold the areas of carbonate deposition fairly constant through time. If the present-day patterns of surface productivity and ratio of deposited to biologically fixed calcite (1:5; Berger, 1970b; Broecker, 1971) have prevailed for the past 35 m.y., then the existence of such a negative feedback is not surprising.

Given the stability of the CCD, it appears that the lysocline and dissolution-rate curves of Figure 61 are responding to changes in the structure of deep water in the equatorial Pacific. Edmond (1974) pointed out that the present lysocline is close to the top of northward-flowing Antarctic Bottom Water in the Pacific. If this relationship has held during the past, Figure 61 indicates that the present regime with a deep abyssal front at about 4 km, which marks the top of a thin layer of corrosive bottom water, has existed only for about the past 15 m.y. From 15 to about 40 m.y. B.P., there is little evidence of marked stratification in the deep water, which suggests that the complex thermohaline structure in today's ocean did not exist, but rather that intermediate and deep waters were well mixed. Whether this indicates a single source of all intermediate and deep waters or several sources with similar salinity-temperature properties is unclear.

Although Figure 61 shows that the Eocene-Oligocene drop in the equatorial CCD is only one of several marked perturbations in the carbonate system during the Cenozoic, the strong impact of this drop on the lithology of the DSDP cores has attracted particular attention (Berger, 1973a; van Andel and Moore, 1974). Our data provide no clue as to the cause of the abrupt decrease in corrosiveness of the bottom water between 37 and 40 m.y. ago. A global budget for calcium carbonate may be required to determine whether the change reflects a shift in the locus of carbonate deposition in the ocean basins or an increase in the rate of supply of dissolved carbonate to the oceans.

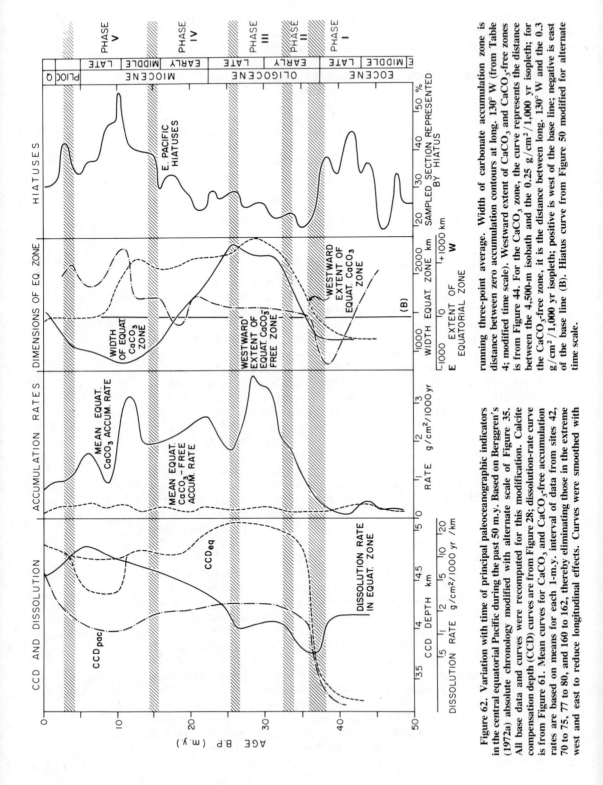

Figure 62. Variation with time of principal paleoceanographic indicators in the central equatorial Pacific during the past 50 m.y. Based on Berggren's (1972a) absolute chronology modified with alternate scale of Figure 35. All base data and curves were recomputed for this modification. Calcite compensation depth (CCD) curves are from Figure 28; dissolution-rate curve is from Figure 61. Mean curves for CaCO₃ and CaCO₃-free accumulation rates are based on means for each 1-m.y. interval of data from sites 42, 70 to 75, 77 to 80, and 160 to 162, thereby eliminating those in the extreme west and east to reduce longitudinal effects. Curves were smoothed with running three-point average. Width of carbonate accumulation zone is distance between zero accumulation contours at long. 130° W (from Table 4; modified time scale). Westward extent of CaCO₃ and CaCO₃-free zones is from Figure 44. For the CaCO₃ zone, the curve represents the distance between the 4,500-m isobath and the 0.25 g/cm²/1,000 yr isopleth; for the CaCO₃-free zone, it is the distance between long. 130° W and the 0.3 g/cm²/1,000 yr isopleth; positive is west of the base line; negative is east of the base line (B). Hiatus curve from Figure 50 modified for alternate time scale.

PALEOCEANOGRAPHY OF THE EQUATORIAL PACIFIC OCEAN

The results of this study have increased our knowledge of the oceanographic evolution of the equatorial Pacific in several principal areas. The first of these is the history of the dissolution of calcium carbonate in deep water. The calcite compensation depth at the Equator dropped about 1,500 m from 40 to 35 m.y. ago (Fig. 62), but it has been remarkably stable since 35 m.y. B.P., ranging from a maximum depth of about 5,100 m in late Oligocene time to a minimum of about 4,400 m during late Miocene time. Yet during this same period, the vertical gradient of the rate of dissolution has varied by a factor of 30—from about 0.6 g/cm²/1,000 yr/km during early Oligocene time to about 18 g/cm²/1,000 yr/km during late Miocene time.

We believe that the gradient of the dissolution rate reacts to changes in the structure of deep and bottom waters, whereas the equatorial CCD is primarily determined by the global carbonate budget and by changes in the locus of carbonate deposition through time (Berger and Winterer, 1974). If these assumptions are correct, there was little stratification of deep waters in the equatorial Pacific during the Oligocene. The present regime, with a thin, well-mixed layer of bottom water beneath a less dense mass of deep water, apparently developed during early Miocene time, perhaps in response to a stable supply of Antarctic Bottom Water. The drop in the equatorial CCD at the end of the Eocene points either to an increase in the rate of supply of dissolved carbonate to the oceans or to an increase in the rate of carbonate deposition in the equatorial Pacific at the expense of other areas of the sea floor. After a very extensive shallow-water (shelf) carbonate build-up, particularly in the form of nummulitic limestone in the belt extending from the East Indies to Gibraltar (which peaked in middle Eocene time), a severe decline took place approximately at the time of the large drop of the CCD and was concomitant with the major development of antarctic glaciation (W. A. Berggren, 1974, written commun.).

We have further shown (preceding section of this chapter) that even after changes in the dissolution of carbonate are taken into account, the variations in the carbonate and carbonate-free accumulation rates since middle Eocene time demand the assumption of changes in the production of biogenic carbonate and silica in the surface water and, hence, imply changes in upwelling and fertility. The exact magnitude of the changes in carbonate supply cannot be defined, because the minimum depositional depths represented in the available drill sites allow calculation of the rate of dissolution as a function of water depth only below about 3 km. In particular, the depth of the lysocline, which in today's ocean corresponds to a marked increase in the rate of dissolution of carbonate, is unknown for the Tertiary. It is clear, however, that the supply of carbonate from surface water prior to late Oligocene time must have been lower than the accumulation rate later (Fig. 61), although the extremely high "extrapolated surface supply" values since early Miocene time probably reflect deepening of the lysocline rather than real increases in the surface biological productivity. The lowest rate of supply (less than 2.5 g/cm²/1,000 yr) occurred about 38 m.y. ago. This low coincides with the striking late Eocene drop of the CCD, with a minimum in the rate of accumulation of noncarbonate debris (Fig. 44), and with the lowest vertical gradients in the carbonate dissolution rate (Fig. 62) observed during the Cenozoic.

The occurrence of numerous unconformities caused by erosion, dissolution, or nondeposition indicates several phases of intensified regional bottom-water flow. In sediments of middle Eocene or younger age, there are two intervals commonly represented by hiatuses: one in late Eocene time (40 to 43 m.y. B.P.) and another in the middle part of the Miocene (10 to 12 m.y. B.P.). The former is coincident with the initial opening of the Tasman seaway (Kennett and others, 1975) to near-surface flow, and the latter with the rapid build-up of the antarctic ice cap (Kennett and others, 1975). The middle Miocene period of peak hiatus abundance (Fig. 62) is unusual, because it occurred during a time of maximum sediment-accumulation rates on the Equator. At all other times, the abundance of hiatuses and the sediment accumulation rates tend to be negatively correlated.

There are also two periods during which hiatuses are relatively uncommon in the equatorial region (Fig. 62). The earlier, middle Eocene minimum is found throughout the Pacific Basin and coincides with a minor peak in accumulation rates, particularly of the carbonate-free fraction. The early Oligocene hiatus minimum is found only in the eastern tropical Pacific, whereas in the western Pacific this interval is represented by an extensive hiatus (Kennett and others, 1972, 1975; Rona, 1973) that appears to be associated with major global cooling and initiation of production of Antarctic Bottom Water (Douglas, 1973; Shackleton and Kennett, 1975a).

The sequence of events discussed above and shown in Figure 62 is a highly generalized version of a much more complex series of fluctuations. These shorter term variations, which end with the well-known high-frequency fluctuations in carbonate content of equatorial cores of Quaternary age (Arrhenius, 1952; Hays and others, 1969), occur on several time scales. Their presence was documented in Chapter 5, but the data, in particular their correlation and chronology, are insufficient to test various hypotheses set forth to account for them.

The past 50 m.y. have witnessed major changes in the global circulation of surface and deep waters of the oceans. These changes resulted from shifts in the configuration of the continents and from a progressive deterioration of climate from an early Cenozoic optimum (Hornibrook, 1968; Kennett and others, 1975); the latter first produced progressive glaciation of Antarctica and, much later, the development of an arctic ice cap (Berggren, 1972b). The information regarding the Cenozoic history of ocean circulation and climate is still fragmentary, but it is improving rapidly with progress in our understanding of plate motion and with continuing progress of the Deep Sea Drilling Project, particularly in the circum-Antarctic region. It is inevitable that in the years immediately ahead, large modifications of present views will be necessary, so that the following attempt to fit paleoceanographic events in the equatorial Pacific into a global ocean evolution is merely an initial exploration certain to be superseded in the near future.

Currently available data allow us to place the oceanographic history of the equatorial Pacific within the context of several interrelated lines of environmental change. Cenozoic shifts in the configuration of continents and oceans have been documented that must have profoundly influenced the surface- and deep-water circulation of all oceans. Of particular interest in the equatorial Pacific is the progressive closure of the circumglobal Tethys seaway, which began in early Oligocene time in the Middle East and was completed in the Mediterranean region in late Oligocene or early Miocene time (Dewey and others, 1973; Phillips and Forsyth, 1972; Pitman and Talwani, 1972). A deep equatorial passage existed north of Australia and south of the Asian block as late as middle Eocene time (Kennett and others, 1972), but it became restricted and then closed in late Eocene and Oligocene time by extensive tectonism along the northern margin of Australia (Veevers, 1969) and in the western Pacific (Moberly, 1972). A final element in the equatorial circumglobal passage—the seaway across the Isthmus of Panama—persisted into late Miocene time but was closed by middle or late Pliocene time (Bandy, 1970; Kaneps, 1970; Malfait and Dinkelman, 1972; Weyl, 1973). On the other hand, whereas circumpolar flow took place around northern Australia as late as early Oligocene time (Kennett and others, 1972), rifting between Australia and Antarctica opened a shallow passage across the South Tasman Rise in early Oligocene time, initiating direct circum-Antarctic surface flow. A deep passage was established in this area in late Oligocene time (Kennett and others, 1975) and permitted development of the full Antarctic deep-current system.

Simultaneously, major climatic changes of a global nature took place. After an initial gradual decline, a sharp and probably global drop in temperature took place near the Eocene-Oligocene boundary, reflected by a drop of about 5° C in sea-surface temperature in the southwestern Pacific (Devereux, 1967; Shackleton and Kennett, 1975a) and a similar but less closely sampled decline in the northwestern Pacific (Douglas and Savin, 1971). Simultaneously, bottom-water temperatures in the southwestern Pacific dropped to about 4° to 5° C, approximately the same temperature as today (Shackleton and Kennett, 1975a). This implies a surface-water temperature near zero in high southern latitudes so that sea-level glaciers and sea ice could develop around Antarctica. However, oxygen-isotope records and the temporal distribution of ice-rafted debris (Geitzenauer and others, 1968; Kennett and Brunner, 1973; Kennett and others, 1975; Margolis and Kennett, 1970, 1971; Shackleton and Kennett, 1975a) indicate that antarctic glaciation did not achieve major dimensions until early middle to late Miocene time, when maximum development of the ice sheet began. A lesser advance took place in early Pliocene time. Major development of northern polar glaciation probably did not occur until about 3 m.y. ago (Berggren, 1972b; Shackleton and Kennett, 1975b). Judging from the rates of biogenic sedimentation, biological productivity in high southern latitudes remained low until early Miocene time, when it gradually began to increase to a maximum in the Pliocene (Kennett and others, 1975). This is supported by the gradual westward advance with time of the equatorial carbonate-free zone (Figs. 44, 62).

In Figure 62 we have summarized the temporal changes of the principal paleoceanographic indicators in the equatorial Pacific during the past 50 m.y. We shall now attempt to correlate this record with the climatic and oceanographic history elsewhere in the world and interpret it in terms of changes in ocean circulation. For this purpose, the equatorial Pacific history can be divided into the following five more or less discrete intervals.

Phase I—Earlier than 38 m.y. B.P. (Middle and Late Eocene)

The interval that occurred before 38 m.y. B.P. was characterized by many hiatuses, shallow CCD, moderately steep vertical gradient of the carbonate dissolution rate, narrow equatorial carbonate zone, very low mean equatorial carbonate accumulation rate, and very limited westward extent of the equatorial carbonate zone.

The most striking feature is the shallow equatorial CCD, which was almost 2 km above its Oligocene depth. Such a dramatic difference cannot be explained by processes restricted to the equatorial Pacific. Rather, it implies either that carbonate deposition during the Eocene was concentrated elsewhere in the oceans (for instance, in the Gibraltar–East Indies belt) or that the rate of input of dissolved carbonate was substantially lower than later during the Cenozoic. The latter hypothesis is consistent with evidence of widespread continental peneplanation during the Eocene, and with Bramlette's (1965) suggestion that Cretaceous-Tertiary faunal changes resulted from a reduced influx of nutrients to the ocean during early Cenozoic time. Confirmation of this suggestion, however, requires reconstruction of global carbonate budgets for early Tertiary time, an undertaking beyond the scope of this study.

The other features of this interval are closely related to the lack of calcareous sediments. The moderately steep vertical dissolution gradient restricted carbonate to a narrow depth zone near the rise crest and close to the Equator. The combination of slow accumulation rates and more porous noncalcareous sediments over most of the area facilitated dissolution and lateral displacement of sediments, leading to the numerous hiatuses of this interval. This abundance of hiatuses and of nonfossiliferous intervals was accentuated by the diagenetic redistribution of opaline silica, which formed the chert found in the western part of the study area.

Phase II—36 to 33 m.y. B.P. (Early Oligocene)

The interval from 36 to 33 m.y. B.P. was characterized by few hiatuses, deep CCD, very shallow vertical gradient of the carbonate dissolution rate, widening equatorial carbonate-rich zone, increasing mean equatorial carbonate accumulation rate, and westward advance of the equatorial carbonate zone.

The transition from the preceding interval is abrupt and is marked by very low noncarbonate supply and carbonate dissolution rates, the lowest during the entire Cenozoic. The overall depression of the CCD points to either a change in locus of carbonate deposition, for example, a shift from shallow continental margins to deep sea, or increased continental weathering. In addition, however, the marked reduction of the vertical gradient of the carbonate dissolution rate, which is responsible for the increased area of carbonate deposition and indirectly for the reduced number of hiatuses and faster mean equatorial accumulation rate, points to a change in the structure of Pacific deep water.

During the Eocene interval, the deep water was apparently somewhat stratified and had a residence time long enough to accumulate oxidative carbon dioxide; this led to the moderate vertical gradient of the carbonate dissolution rate. During early Oligocene time, however, the residence time apparently decreased markedly, probably in combination with a breakdown of the stratification, which produced a very gentle gradient of the dissolution rate. In this regard, this interval resembles glacial conditions of the Pleistocene.

The cause for the change in bottom-water characteristics is uncertain, but the extension of antarctic glaciers to sea level (Kennett and others, 1975; Shackleton and Kennett, 1975a), together with the first opening of the shallow passage across the South Tasman Rise between Australia and Antarctica, may have sharply increased the rate of formation of bottom water. This increase thereby reduced the corrosiveness of the bottom water (less time for oxidative CO_2 build-up) and broke down the stratification of the Eocene deep water. The best estimate for the time of these events is about 37 m.y. B.P. (Kennett and others, 1975).

Phase III—32 to 26 m.y. B.P. (Early to Late Oligocene)

The interval from 32 to 26 m.y. B.P. was characterized by few hiatuses, very deep CCD, moderately steep vertical gradient of the carbonate dissolution rate, very wide equatorial carbonate-rich zone, very high mean equatorial carbonate accumulation rate, equatorial carbonate-rich zone extending farthest west, and equatorial carbonate-free zone advancing westward.

The transition from phase II to phase III was gradational, the principal feature being an increase in the gradient of the dissolution rate. The effects of this increase were apparently nullified, however, by an increase in the carbonate supply rate or by a depression of the lysocline, which led to the deepest equatorial CCD, widest and westernmost equatorial carbonate zone, and highest equatorial carbonate accumulation rate recorded during the entire Cenozoic.

The steepening of the dissolution-rate gradient during the early Oligocene points to a decrease in the influx of deep water, possibly resulting from changes in hydrography around Antarctica that developed as the passage across the South Tasman Rise deepened and provided increasing room for deeper circum-Antarctic circulation (Kennett and others, 1975). The gradual closing of the Tethys seaway during this same interval may have been responsible for a narrowing of the Pacific equatorial current system (Luyendyk and others, 1972), with accompanying intensification of the upwelling that supplied the carbonate responsible for the depressed CCD and the high accumulation rates.

Phase IV—26 to 15 m.y. B.P. (Late Oligocene to Early Miocene)

The interval from 26 to 15 m.y. B.P. was a fairly stable but transitional one with the following characteristics: few but increasing hiatuses, shoaling CCD, steepening vertical gradient of the carbonate-dissolution rate, decreasing width of the equatorial carbonate-rich zone, moderate mean equatorial carbonate accumulation rate, and a stationary western limit of the equatorial carbonate-rich zone.

These features suggest that the influx of deep and bottom waters decreased throughout the interval, resulting in CO_2 build-up and greater dissolution of carbonate, with consequent shoaling of the CCD, narrowing of the equatorial carbonate zone, and increase in the occurrence of hiatuses. The relationship between the equatorial CCD and the gradient of the carbonate dissolution rate (Fig. 61) indicates that the lysocline became established close to its present position by 15 m.y. B.P. This suggests that the complex of source areas and mechanisms that produces the stratified intermediate and deep waters of the Pacific today approached its present configuration and intensity at that time. This is in accord with the major development of antarctic glaciation and the evidence for the full establishment of the Antarctic current system and convergence noted by Kennett and others (1975). A rather marked reduction in the westward extension of the carbonate-free equatorial zone (Fig. 62) may possibly indicate a temporary warming trend with reduced fertility (Kennett and others, 1975; Shackleton and Kennett, 1975a). A similar reduction around 42 m.y. B.P. may be due to the presence of unrecognized hiatuses, but this explanation is not likely for the later event.

Phase V—15 to 4 m.y. (Middle Miocene to Early Pliocene)

The interval from 15 to 4 m.y. B.P. was characterized by many hiatuses, shallower CCD, steep vertical gradient of the carbonate dissolution rate, narrow equatorial carbonate-rich zone, variable mean equatorial carbonate accumulation rate, large but temporary westward advance of equatorial carbonate-free zone, and reduction in westward extent of equatorial carbonate-rich zone.

The transition from phase IV to phase V is generally gradational, so that the 15-m.y. B.P. boundary is largely based on an abrupt increase in the number of hiatuses in the eastern Pacific (Fig. 62). As mentioned previously, the thermal structure of the oceans during this interval appears to have been very similar to the pattern today. This may reflect increased antarctic glaciation (Kennett and others, 1975), which produced the first true Antarctic Bottom Water about 13 to 15 m.y. ago. The development of a thin, corrosive layer of bottom water would explain the shoaling of the CCD as well as the steep dissolution-rate gradient, narrow equatorial carbonate zone, and abundant hiatuses.

The fluctuations in the mean equatorial carbonate accumulation rate are puzzling. The peak at 11 to 12 m.y. B.P. is particularly enigmatic because it violates the correlations between the various parameters illustrated in Figure 62 that otherwise prevail throughout the entire Cenozoic. One possible explanation is that the northward migration of Australia cut the Pacific-Indian equatorial current system at this time, resulting in a pile-up of surface water in the western Pacific and development of the easterly flowing Cromwell Current. Upwelling associated with the Cromwell Current could have produced a narrow zone of rapid carbonate deposition along the Equator. Some support for this suggestion comes from the curve of the carbonate-free accumulation rate, which also has a peak at 11 m.y. B.P., and from the large westward advance at this time of the equatorial carbonate-free zone.

Alternatively, the peak at 11 to 12 m.y. B.P. and the low at 9 m.y. B.P. in the carbonate accumulation rate could result from inaccuracies in the assignment of absolute ages to this interval. As already mentioned, errors in the duration of the 1-m.y. intervals are reflected directly in accumulation-rate estimates. We discussed in Chapter 5 the uncertainties that plague the time scale for middle Miocene time (see also Moore, 1972). Even though the peak at 11 to 12 m.y. B.P. appears in accumulation-rate curves based on both the Berggren (1972a) and modified time scales used in this study, the possibility that it is an artifact introduced during the conversion from biostratigraphic to absolute ages cannot be completely discounted.

The end of phase V marks an expansion of antarctic glaciation (Kennett and others, 1975) and just precedes the onset of arctic glaciation (Berggren, 1972b). During the past 4 to 5 m.y., variations in biogenic sedimentation in the equatorial Pacific have been dominated by glacial-interglacial fluctuations (see, for example, Arrhenius, 1952; Berger, 1973c; Hays and others, 1969; Pisias, 1974). The curves in Figure 62 show an attempt to average these fluctuations, but the trends must be considered tentative because of the small number of samples and inadequate resolution of the time scale.

Much of the discussion above is tentative and is subject to uncertainties of correlation and interpretation. However, it is obvious that the developing glacial climates, especially in the region of the South Pole, and the changing configuration of continents and oceans have strongly influenced, and to a large extent controlled, the oceanographic evolution of the Pacific equatorial region as we have been able to trace it. In fact, the events in this region appear to be so closely related to the climatic and oceanographic history of Antarctica that this area can be considered the motor that drove and drives the equatorial Pacific depositional and paleoceanographic evolution.

8

Paleoceanographic Studies:
Problems and Prospects

*Les dépots des grands fonds océaniques représentent des archives
complètes, sans lacunes, de l'histoire de la terre.*

J. Bourcart, 1951

The comfortable view expressed by Bourcart has long since been abandoned, but the archives, although full of gaps, have proved to be fascinating. The deep-sea drilling records constitute the best access to these archives, and primarily on the basis of published DSDP records, we designed the present study, among other things, to explore the methodology and potential of paleoceanographic studies. In the process we hoped to develop some effective procedures for dealing with the data, to acquire insight into the potential of the DSDP material for such purposes, and to stimulate interest in such investigations. Finally, we wished to focus as sharply as possible on the variety of problems involved in paleoceanographic studies based on pre-Quaternary deposits and to develop some sense of priorities. At the end of a long and sometimes rather arduous task, we are content with the degree to which we have achieved these objectives, although frustrated by the vast amount of information we have not been able to touch upon and, to some extent, by the difficult task of casting published DSDP data into usable form. In this last chapter, we reflect on some of the more general aspects of our exploration, on the requirements of this type of investigation, on what it has taught us procedurally, and on perspectives for the future.

Several things stand out clearly in retrospect: the importance of quantitative information, especially on sedimentation rates; the critical role of a high-resolution absolute chronology and of precise plate-migration data; and the impossibility of interpreting paleoceanographic indicators from a single, albeit large and important, region without equivalent global information.

The key to the results obtained in this study is the use of quantitative age, composition, and accumulation-rate data. The impact of this is clear when we compare the discussion in Chapter 3 or that of Tracey and others (1971), Hays and others (1972), or van Andel and Heath (1973b) with that in Chapter 7. Only absolute chronology, combined with quantitative lithologic, paleontologic, and petrographic data, permits the use of rate-of-change information and only rate information permits paleoceanographic inference. We admit to an initial scepticism regarding the feasibility of obtaining from the *Initial Reports* quantitative information of adequate quality to permit this approach, and we did experience considerable difficulty in segregating acceptable from unacceptable data. The internal consistency of the results, however, demonstrates to our satisfaction that the approach has been successful and that it holds promise for the future. Nevertheless, there is ample room for improvement. Improvements in the quality of the DSDP data—in particular, the density and porosity information (which was much more crucial for sedimentation-rate calculations than originally envisaged) and the x-ray mineralogic data (which was not quantifiable and, therefore, proved useless)—could be fairly easily achieved. Core quality and core recovery also are amenable to considerable improvement, partly by improved technology but also by better recognition that sparse coring, no matter how efficient within a narrow definition of the objectives of the site, seriously damages its use for most purposes. Moore (1972) compiled statistics that show that core recovery not only varies widely but is totally inadequate at a very large number of sites. In this respect, the central Pacific Ocean is not representative; the coring record there is far better than average, perhaps because sedimentologists rather than plate tectonicists led the cruises. In other areas—in particular the Atlantic, where both site selection and coring programs were dominated by tectonic objectives—the data will be of much less value for the reconstruction of the depositional and paleoceanographic history of the region. Core distortion and large and unknown coring gaps between adjacent cores are other serious problems that can be remedied only by advances in technology. In the present case, these have not proved to be too much of an impediment, but they are part of the reason why the nature and time dependence of high-frequency events have eluded us, and they limit severely our ability to recognize hiatuses of short duration.

Vast paleoecological and evolutionary information is contained in the paleontologic records. This includes both species composition and preservation phenomena. The lack of quantitative data, unfortunately, has compelled us to abandon this source; the data in the *Initial Reports* are useful mainly for biostratigraphic purposes. Studies of the original samples are needed to tap this reservoir, and important information can be expected from them. This task is enormous, however, and also very time-consuming, and such data may never be routinely available.

The most critical component in our study is the absolute chronology (Chap. 5; App. 1). We have concluded from somewhat circumstantial evidence that the chronology of Berggren (1972a) is usable, although improvements are required, particularly for the middle part of the Miocene. Some clear limitations that will constrain future research have also become apparent. The first of these we have already encountered. It is clear from previous chapters that a temporal resolution on the order of 1 or 2 m.y. has been the key to resolving most of the events we have discussed. Lower resolution would have essentially stymied the attempt at unraveling the depositional history of the equatorial Pacific. Two problems are involved: the ability to resolve two events spaced no more than 1 or 2 m.y. apart and the ability to correlate such events regionally. With regard to the first problem, the good core recovery and the generally adequate preservation of faunal assemblages make such resolution feasible in the equatorial Pacific. Elsewhere, this is often not the case. Regarding the second problem, within most regions such correlations during the Cenozoic can usually be made, but correlations between high and low latitudes are more controversial and less precise. This has caused some difficulty in establishing the synchroneity of events in the equatorial Pacific with those in the circum-Antarctic region and will become a more serious problem when attempts are made to correlate events on a global scale.

For earliest Cenozoic time and in particular for the Cretaceous, the situation is entirely different. Given present biostratigraphic knowledge (Baldwin and others,

1974), it appears impossible to extend this type of investigation far beyond the age limits of the present study; unless the rate of change of processes prior to 50 m.y. B.P. was much slower than it was subsequently (an improbable assumption), we seriously doubt whether real insight into the depositional history of the early Cenozoic and Cretaceous oceans can be obtained at this time. A large amount of isotopic and biostratigraphic work is required to make such investigations really attractive.

Finally, high-frequency variations in depositional conditions will lend themselves to real interpretation only when it becomes possible to establish accumulation and solution rates and when the events can be precisely correlated not only between sites but also between regions. The degree of resolution and accuracy of correlation that is required is well beyond the present state of the art.

Elsewhere in this study we have emphasized the critical nature of the paleotectonic base, particularly of two aspects: the determination of paleolatitude and paleolongitude and the determination of subsidence histories of the drill sites. The equatorial Pacific Ocean is particularly favorable in both regards (and was in fact chosen partly for that reason), because the data points are dominantly located on a single plate with a fairly simple rotational history that could be reconstructed from two independent lines of evidence. Such a situation is exceptional, and the reconstruction of paleolatitude and paleolongitude in regions such as the Indian Ocean or western Pacific Ocean will encounter much greater difficulties and involve greater final uncertainties. As we have shown in Chapter 4, an uncertainty in the paleolatitude of only a few degrees can alter the results of the sedimentation study significantly. The central Pacific Ocean is also favored because the subsidence history can be inferred rather simply from the subsidence of a single oceanic rise. In tectonically more complex regions, such as the western and southwestern Pacific, the reconstruction of paleodepths to a precision of a few hundred metres will be much more difficult and sometimes impossible. This is especially unfortunate for one specific and critical line of investigation—the temporal variation of the depth of the lysocline. We have concluded that temporal variations of fertility, and hence of upwelling and surface circulation, can be effectively approached by resolving simultaneously the histories of the lysocline and of the vertical gradient of the carbonate dissolution rate (Chap. 7). In fact, the only other avenues to the study of surface circulation patterns of the past are the paleoecologic analysis of faunal and floral assemblages and the measurement of the accumulation rates of siliceous biogenic material. The former technique, although highly successful in studies of latest Quaternary time, is time consuming and will entail great technical difficulties for earlier Cenozoic time, whereas the latter technique is still unproven. Thus, the lysocline approach is important. This, however, requires the study of drill sites that include the lysocline in their paleodepth range. In general, paleodepths associated with mid-ocean rises are too deep, and the study must be based on sites in anomalously shallow locations such as the Hess and Manihiki Rises and the Carnegie Ridge. Because of the anomalous depth of these locations, their subsidence histories cannot be determined solely from simple age-depth relationships.

Given the vast amount of data involved in this study, computer data processing is clearly necessary. With the aid of machine data handling, for example, we could have generated isopleth maps for all variables at 1-m.y. intervals rather than for a select few, thereby improving the resolution of curves such as those in Figure 62. For this study, the gradual evolution of our data set made this impractical, but in the future, this and other computational techniques should accelerate the work. However, there are some important difficulties. The large number of possible environmental factors relative to the restricted size of the data matrix renders statistical analysis inefficient. Furthermore, although we have based most of our conclusions on time sequences of variables, we have avoided more sophisticated time-series analysis techniques. This was not by choice but because of a theoretical deficiency, namely the lack of theory for time series in which uncertainties of estimate are attached to both the time and dependent variables. Such a theory would have appreciably sharpened our definition of the errors involved and should receive immediate attention.

In this paper we have attempted an essentially kinematic description of the paleoceanography of the equatorial Pacific Ocean, and in a highly generalized form at that. This description has been achieved by postulating causal relationships between events for which, in reality, only synchroneity can be demonstrated and by using analogies from modern ocean dynamics as a means to select appropriate pairs of events from the many possible ones. Apart from the difficulties inherent in establishing synchroneity, the approach is obviously in danger of extensive circular reasoning. In addition, one can reasonably argue that the degree of generalization may be such that it virtually prevents inferences regarding the dynamic processes underlying the observed events because of critical discrepancies in time scales. An example is the history of the past 4 to 5 m.y. depicted in Figure 62; clearly, the generalized history made possible by the limited data is inadequate to unravel the complex events of the alternating glacial and interglacial periods. The study of rates of change of certain paleoceanographic indicators may possibly add significant insight, but better and more closely spaced data would be required. A case in point is the difference between the rates of change since the Eocene-Oligocene boundary of the width and accumulation rate of the equatorial zone on one hand and the carbonate dissolution rate on the other.

The same inevitable generalization of the histories of the various indicators also renders it unlikely that more quantitative approaches, such as the construction of numercial dynamic models to fit the observations, will be as applicable as it promises to be in late Quaternary paleoceanography. Perhaps in the future, when some global patterns begin to emerge, the construction of generalized models may be helpful in deducing the underlying dynamic evolution of the oceans.

It is evident that the results of this study do not represent a comprehenisve Cenozoic paleoceanography of the equatorial Pacific; in fact, they are at best a mere beginning. Nevertheless, several useful lines of research necessary to carry this and similar studies forward can now be rather sharply delineated. Some of these we have mentioned above: improved biostratigraphy and high-resolution absolute chronology, including better correlations across latitudinal zones, and comprehensive reconstruction of the plate tectonic history of the oceans. Given the present state of both fields, we believe that major advances are possible by investigating the paleoceanography of well-chosen regions elsewhere in the world along lines similar to those used here, suitably modified to local conditions. The paleoceanography of the Atlantic Ocean, although likely to be restricted by poor drill-site location and inadequate core recovery, is needed to elucidate the history of development of the North Atlantic Intermediate and Deep Waters and the inflow of Atlantic Intermediate Water into the Pacific, with its consequences for the Pacific thermohaline circulation. Furthermore, it is possible that such a study would clarify the history of surface circulation through the Tethys seaway and the Panamanian isthmus. Obviously, further delineation of the circum-Antarctic history is of the greatest importance; the results of Kennett and others (1975) are highly promising, and the completion of the circum-Antarctic drilling program should allow a comprehensive and quantitative synthesis. Together, these two regions provide essential data for the interpretation of the Pacific equatorial circulation.

It is also obvious that the reconstruction of past surface-circulation patterns is of such importance that it must be pursued by all possible means, notwithstanding our initial lack of success in addressing the problem.

In conclusion, we find ourselves encouraged by this approach. It is obvious that many fruitful lines of research are made possible by the DSDP material, either in the form of data that are already available or data that can be generated by analysis of samples, and that a regional approach to the use of these resources is productive. It seems to us that the experience gained in this experiment has produced some unexpected ways of viewing the data, some unexpected results, and also some unexpected problems, limitations, and difficulties. With the high hopes we have for the knowledge contained in the data resources, we conclude that for a long time to come it will remain a territory of scientific adventure, requiring new trials, new methods, and new creative thinking, and that a highly structured and consolidated approach is as yet neither possible nor desirable.

Appendix 1

Data Sources and Data Processing

INTRODUCTION

Most of the stratigraphic and lithologic information used in this study comes from approximately 20 drill sites occupied on Legs 5 (sites 40 to 42), 8 (sites 69 to 75), 9 (sites 77 to 83), and 16 (sites 159 to 163) of the Deep Sea Drilling Project (Fig. 63). The data for these sites have been reported by McManus and Burns (1970), Tracey and Sutton (1971), Hays (1972), and van Andel and Heath (1973a). Information from other drill sites in the area (Winterer and Riedel, 1971; Winterer and Ewing, 1973) has been used where appropriate but has not been processed in the same comprehensive manner.

The drilling campaigns span the interval from April 1969 to May 1971. Not surprisingly, large changes in reporting formats, shipboard and shore laboratory procedures, and biostratigraphic zonations and nomenclature took place during this two-year period early in the history of the Deep Sea Drilling Project. As a result, a substantial effort was required to render the data comparable in quality and format from leg to leg and site to site and to establish a uniform biostratigraphy.

In general, the data reported in the *Initial Reports* could be converted directly to the adopted standard formats, but in some cases data gaps, errors, or the low resolution of graphic displays required extraction of the information from prime-data files of the Deep Sea Drilling Project or additional analytical work on samples. In the *Initial Reports*, errors are common, and much care was required to eliminate these as much as possible from the data body used in this study and presented in Appendixes 2 to 4.

Information from surface cores has been used in addition to that of the drill sites (Fig. 63). These surface cores, which penetrate pre-Quaternary sedimentary deposits, are from the collections of Scripps Institution of Oceanography and Lamont-Doherty Geological Observatory. The biostratigraphic age assignments were either made especially for this study by one of us (Scripps cores) or came from Saito and others (1974). The data have been tabulated in Appendix 5.

CORE POSITION IN HOLE

The most basic parameter, on which all others depend to some extent, is the relationship between age and depth in the hole. Thus, its accuracy is of paramount importance. Even in the case of continuous coring, the depth of any individual sample cannot be determined to better than ±1 or 2 m, because this amount of loss or gain between coring trips is common. For a low sedimentation rate, this can mean an age uncertainty of 1 m.y. We have accepted the core and final depths given in the drilling statistics tables in the site chapters of the *Initial Reports*. Some cores exceed the cored interval in length; in such cases, we have adjusted the core length to the cored interval.

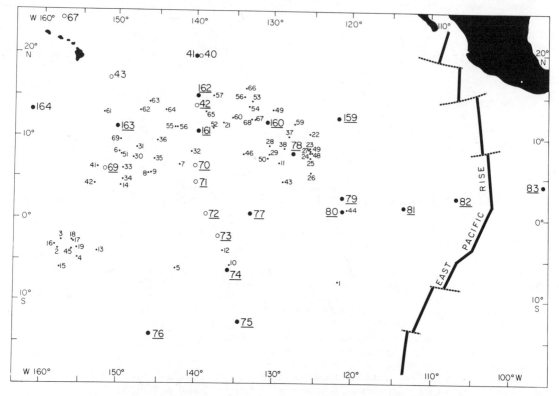

Figure 63. Location of drill sites and surface cores. Large black dots: drill sites penetrating basement; open circles: drill sites terminating in sediment. Underlined site numbers indicate sites tabulated in Appendixes 3 and 4. Small black dots are surface cores containing pre-Quaternary deposits; numbers refer to Appendix 5, where actual core numbers are given.

When recovery is incomplete, assignment of the core section to the cored interval is arbitrary, and conventions vary from assignment to the top of the interval in some *Initial Reports* to routine assignment to the bottom in others. To facilitate comparison with the *Initial Reports*, we have followed the convention used in each. Sometimes the position assigned in the physical-properties log is at variance with the one used in the graphic core description. In those cases, the physical-properties log positions have been adjusted.

BIOSTRATIGRAPHY AND ABSOLUTE CHRONOLOGY

In stratigraphic studies of marine sediment, ages are frequently given in millions of years rather than in terms of zone names, as has been customary in land geology. This practice has many advantages as long as its limitations are clearly understood. It provides a simple and linear time scale that can be finely interpolated if combinations of foraminiferal, radiolarian, and nannofossil zones are used. It avoids the use of a cumbersome and specialized stratigraphic nomenclature. It allows direct computation of rates of deposition. Finally, it permits comparison with tectonic events that are usually defined in terms of the geomagnetic polarity reversal time scale or dated by means of isotopic ages of igneous rocks.

The absolute time scale used is derived from Berggren's (1969b, 1972a) calibration of the zonation of planktonic foraminifers with isotopic ages. The absolute chronologies used in conjunction with radiolarian (Moore, 1971; Dinkelman, 1973) and calcareous nannofossil stratigraphies (Bukry, 1973) were derived from biostratigraphic correlation with Berggren's zonation. They do not add absolute-age information, but they do interpolate a finer scale and carry the absolute scale to deposits not containing planktonic foraminifers. The Berggren calibration rests on a relatively small number of isotopic ages connected with the foraminiferal zones by sometimes complex correlations, a fact occasionally forgotten in the proliferation of zone boundaries with attached absolute ages.

The errors in the absolute ages of the zone boundaries cannot be rigorously estimated. Instead, van Andel and Bukry (1973) assumed that the error increases exponentially with age according to the function B.U. = K(age)^c, where B.U. is the zone boundary uncertainty, and K and c are constants. For assumed uncertainties of ±1 m.y. at age 10 m.y. B.P. and ±5 m.y. at age 100 m.y. B.P., K and c take the values of 0.2 and 0.7 (Fig. 64). We have used the relationship in Figure 64 to estimate the confidence limits for the correlation of any two age levels in two drill sites. The error does not exceed 1 m.y. for deposits less than 10 m.y. old, 2 m.y. for deposits between 10 and 27 m.y. old, and 3 m.y. for sedimentary deposits with ages between 27 and 50 m.y. The magnitude of this error at each point is constrained by the points on either side. The implicit assumption is, of course, that the Berggren absolute chronology does not suffer from systematic distortions but only from random errors. The effect of possible systematic errors is discussed in Chapter 5.

Zone definitions and zone boundaries vary from report to report. Commonly, the foraminiferal stratigraphy is based fairly directly on Berggren (1969b, 1972a); only for Leg 9 (Hays, 1972) was it necessary to revise the zonation. The radiolarian stratigraphies are generally based

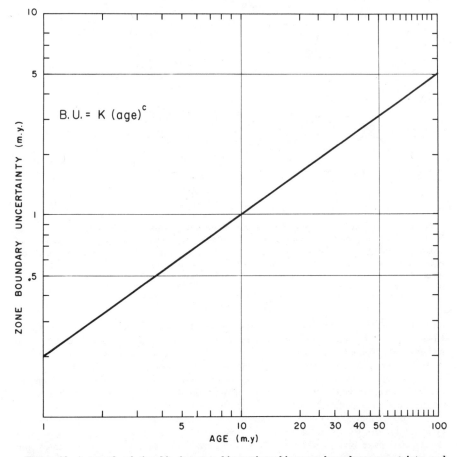

Figure 64. Assumed relationship between biostratigraphic zone boundary uncertainty and absolute age used for determination of absolute-age confidence limits. Modified after van Andel and Bukry (1973, Fig. 2).

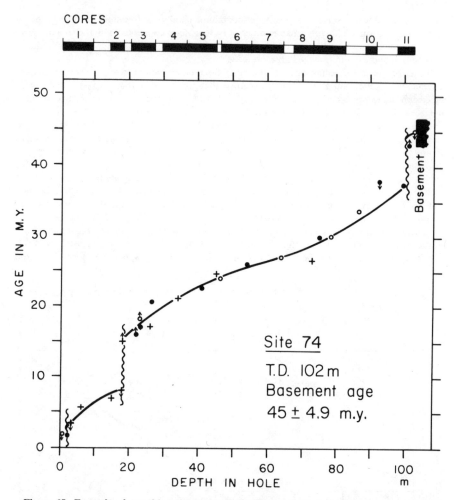

Figure 65. Example of graphic method for determining depth in hole at 1-m.y. intervals. Black dots: foraminiferal zone boundaries; open circles: calcareous nannofossil zone boundaries; crosses: radiolarian zone boundaries. Up and down arrows indicate minimum and maximum ages, respectively. Biostratigraphic and absolute-age data from Appendix 4. Wavy lines are assumed unconformities. Position and number of cores are indicated at top. Smooth curve fitted by hand.

on Moore (1971), with revisions by Dinkelman (1973), but for sites 40, 41, and 42 of Leg 5 (see App. 2) a complete re-examination of original samples was necessary. The calcareous nannofossil stratigraphy of Leg 16 (Bukry, 1973) was used as reported; for all other drill sites, D. Bukry completely revised the nannofossil stratigraphy by means of a re-examination of the original samples (see App. 4). The biostratigraphic zonations and the assignments of absolute age for all drill sites are given in Appendix 4, except for sites 40 and 42, which can be found in Appendix 2.

The basic absolute chronology is from Berggren (1972a). For the ages of radiolarian zone boundaries above the *Calocycletta virginis* zone, data from Moore (1971) was used, whereas the correlations of Bukry and others (1973) were used below this zone and for the calcareous nannofossil stratigraphy. Ages of sediment-basement contacts are from van Andel and Bukry (1973). Their method of minimum overlap was also used to determine the age of the oldest cored deposit in holes that did not reach basement.

For each drill site, the ages of all available zone boundaries were plotted against depth in the hole (Fig. 65). Where the zone boundary was not defined to within a few metres, a maximum or minimum age was substituted. Moore (1972) showed that zone boundaries are not randomly distributed throughout cores but cluster at the top and bottom; in the absence of a good way to evaluate the extent of this distortion, the boundaries were plotted where they had been identified, and a smooth curve was fitted by hand to all points. This curve was used to obtain the depth in the hole of each 1-m.y. boundary. Unconformities were postulated where the depth-age curve approached the vertical (rate of deposition below 1 m/m.y.) or where other evidence—such as missing zones or parts of zones, reworked fossils, or abrupt lithologic changes—was recorded in the *Initial Reports*. Unconformities of less than 1-m.y. duration have been disregarded unless they were firmly established by independent evidence.

The scatter of zone-boundary ages in Figure 65 is representative; it is generally on the order of 0.5 to 1.0 m.y. with occasionally larger excursions. In part, this scatter is due to imperfect data: inadequate sample coverage, poor fossil preservation, and errors in depth assignment of cores and core sections. In part, it may reflect true short-term fluctuations of the sedimentation rate that cannot be assessed with the resolution of the absolute time scale. The curve smoothing is a reasonable way to eliminate these two principal effects.

Undoubtedly, in the future there will be refinements and improvements to the Berggren (1972a) time scale. An example is the alternate time scale for the middle part of the Miocene described in Chapter 5 (Fig. 35). Such changes do not necessarily require complete recalculation of the sedimentation and accumulation rates presented in Appendix 3. As long as the changes are not too large, a table can be constructed that relates the new time scale to the old one, and correction factors can be calculated that will directly convert the tabulated rates to the new scale. The alternate time scale of Figure 35 provides an example of such a conversion. Given the two time scales (Fig. 35, bottom), time equivalences can be constructed.

Alternate scale (m.y. B.P.)	Corresponding Berggren scale (m.y. B.P.)
10 to 11	10 to 13
11 to 12	13 to 14
12 to 13	14 to 14.4
13 to 14	14.4 to 14.8
14 to 15	14.8 to 15.3
15 to 16	15.3 to 16.0

From these equivalences, conversion factors can be devised. As an example, the sedimentation rates for site 77 are converted in Table 5, and converted data for all sites are given in Table 6.

TABLE 5. EXAMPLE OF CONVERSION OF RATE DATA BASED ON BERGGREN TIME SCALE TO ALTERNATE TIME SCALE FOR 10 TO 16 M.Y. B.P.

Berggren scale (m.y.)	Sedimentation rate (m/m.y.)		Sedimentation rate (m/m.y.)	Alternate scale (m.y.)
10–11	4 ⎫			
11				
11–12	5 ⎬ sum		17	10–11
12–13	8 ⎭			
13–14	22		22	11–12
14–15	40 (40 × 0.4)		16	12–13
	(40 × 0.4)		16	13–14
	(40 × 0.2) ⎫	sum	22	14–15
15–16	13 (13 × 0.3) ⎭			
	(13 × 0.7)		9	15–16

LITHOLOGIC CLASSIFICATION

The lithologic classifications used in the *Initial Reports* rest on visual core description and on microscopic examination of smear slides. The quality and amount of detail of the descriptions vary widely from cruise leg to cruise leg and even from site to site. Each of the *Initial Reports*, moreover, uses its own lithologic classification, usually a scheme that is rather similar to that of Arrhenius (1952) or Olausson (1960) but varying considerably in class boundaries, terminology, and genetic emphasis.

The present study relies primarily on quantitative parameters such as calcium carbonate content and rate of deposition and requires only a summary descriptive classification of the lithology. The system adopted to standardize the reported data was modified from that of Olausson (1960) and is very similar to the standard classification recently adopted by the Deep Sea Drilling Project (1974, DSDP Shipboard Manual). It is a strictly descriptive scheme, based only on shipboard data, although the carbonate-content limits have been checked against laboratory determinations.

The following sediment types are recognized:

Brown clay (pelagic clay). Brown clay with common to abundant authigenic components (zeolites, Fe-Mn micronodules), fish debris, and other indicators of very slow deposition. No calcium carbonate is present, and radiolarians and other identifiable siliceous skeletons account for less than 30 percent of the clay. It is called *radiolarian clay* when radiolarian

TABLE 6. SEDIMENTATION AND ACCUMULATION

Age (m.y. B.P.)	a	b	c	d	a	b	c	d
	Site 69				Site 70			
10–11	14.5	0.41	0.04	0.38	10	0.63	0.32	0.31
11–12	5.5	0.24	0.05	0.19	4	0.29	0.12	0.17
12–13	2.2	0.11	0.02	0.09	2	0.12	0.07	0.05
13–14	2.2	0.11	0.02	0.09	1.5	0.12	0.07	0.05
14–15	2.7	0.14	0.01	0.13	2	0.17	0.11	0.06
15–16	3.3	0.16	0.01	0.15	3	0.21	0.12	0.09
	Site 71				Site 72			
10–11	30	2.56	1.96	0.60	30	3.4	2.57	0.57
11–12	50	5.97	5.19	0.78	14	1.52	1.29	0.23
12–13	18	2.24	1.93	0.31	13	1.37	1.18	0.19
13–14	18	2.24	1.93	0.31	12	1.37	1.18	0.19
14–15	25	2.32	1.83	0.49	17	1.80	1.63	0.17
15–16	21	2.41	1.94	0.47	21	2.45	2.09	0.36
	Site 73				Site 74			
10–11	1	0.04	0.01	0.03	..	..	..	..
11–12	9	0.94	0.76	0.18	..	..	..	..
12–13	4	0.50	0.41	0.09	..	..	..	..
13–14	4	0.50	0.41	0.09	..	..	..	..
14–15	5	0.69	0.55	0.14	1	0.03	0.00	0.03
15–16	7	0.80	0.68	0.12	1	0.05	0.00	0.05
	Site 75				Site 77			
10–11	..	..	..	..	17	1.26	0.95	0.31
11–12	..	..	..	..	22	2.19	1.97	0.24
12–13	..	..	..	..	16	1.59	1.39	0.20
13–14	..	..	..	..	16	1.59	1.39	0.20
14–15	1	0.04	0.02	0.02	12	1.27	1.05	0.22
15–16	1	0.06	0.05	0.01	9	0.94	0.73	0.21

remains exceed 10 percent in smear-slide description. This sediment used to be called "red clay."

A red or brown clay enriched in metal oxides is sometimes found immediately above basement; similar enrichment may also affect calcareous sediment. This sediment type, although sometimes superficially similar to brown clay of pelagic origin and sometimes so described in the *Initial Reports*, is of a wholly different origin and is formed by hydrothermal processes (Cronan and others, 1972; Dymond and others, 1973). This type is indicated with an overprinted symbol.

Radiolarian ooze (*radiolarite* when indurated). Dominantly consisting of siliceous (opaline) skeletal material with less than 30 percent calcium carbonate. When containing 10 to 30 percent $CaCO_3$, the sediment is called *nannofossil* or *foraminiferal radiolarian ooze,* depending on the dominant calcareous component.

Nannofossil ooze (*chalk* when indurated). More than 30 percent $CaCO_3$ and less than 20 percent foraminifers in smear slide. If smear slide contains more than 10 percent radiolarians, it is a *radiolarian nannofossil ooze,* and if there is more than 20 percent foraminifers it is *foraminiferal radiolarian ooze.* If clay content exceeds 30 percent, the sediment is called a *marl.*

Limestone (more than 30 percent $CaCO_3$, strongly cemented) and *chert* are separately recognized.

For the construction of the graphic lithologic logs of Appendix 3, cores were plotted against the absolute-age scale, and lithologies were interpolated.

RATES FOR THE INTERVAL 10 TO 16 M.Y. B.P.

Age (m.y. B.P.)	a	b	c	d	a	b	c	d
		Site 78				Site 79		
10–11	7	0.57	0.49	0.08	41	2.49	··	··
11–12	7	0.56	0.49	0.07	38	2.67	··	··
12–13	3	0.22	0.19	0.03	16	1.21	··	··
13–14	3	0.22	0.19	0.03	16	1.21	··	··
14–15	3	0.28	0.23	0.05	21	1.57	··	··
15–16	5	0.34	0.28	0.06	25	1.91	··	··
		Site 80				Site 81		
10–11					67	··	··	··
11–12	7	0.39	0.34	0.05	31	··	··	··
12–13	7	0.61	0.49	0.12	19	··	··	··
13–14	7	0.61	0.49	0.12	19	··	··	··
14–15	11	0.93	0.78	0.15	26	··	··	··
15–16	14	1.26	1.04	0.22	32	··	··	··
		Site 159						
10–11	3.5	0.22	0.00	0.22				
11–12	4	0.24	0.01	0.23				
12–13	2	0.10	0.01	0.09				
13–14	2	0.10	0.02	0.08				
14–15	2.5	0.13	0.02	0.11				
15–16	3.5	0.15	0.03	0.12				

Note: Rates are recalculated on the basis of the alternate time scale of Figure 35. Column headings are as follows: a, sedimentation rate in metres per million years; b, bulk accumulation rate in $g/cm^2/1,000$ yr; c, carbonate accumulation rate in $g/cm^2/1,000$ yr; d, carbonate-free (residual) accumulation rate in $g/cm^2/1,000$ yr. Conversion data in Appendix 3.

CALCIUM CARBONATE CONTENT

Calcium carbonate determinations on one or two samples per core section are routinely made by the shore laboratory of the Deep Sea Drilling Project. The determinations are made with a LECO Carbon Analyzer[R] according to procedures that have varied slightly from leg to leg. The methods have been described by Bader (1970) and Boyce and Bode (1973); revisions are noted in Bode and Cronan (1973). Carbonate determinations for site 159, Leg 16, were made at Oregon State University by a comparable method. Standard deviations of the $CaCO_3$ values determined by the Deep Sea Drilling Project are given by Bode and Cronan (1973) as ±2 percent for the range 10 to 100 percent $CaCO_3$ and 0.6 percent for the range 0 to 10 percent. The OSU determinations have similar standard deviations.

$CaCO_3$ values plotted against depth in the hole were obtained from a computer output of prime data, because the graphs in the *Initial Reports* are very small and sometimes in error. On these computer plots a smooth curve was fitted by hand to all data points and digitized at 2-m intervals beginning at the sea floor. The smooth curve was carried across small data gaps but was interrupted for gaps more than 5 m long. The digitized values were used to compute mean $CaCO_3$ percentages for each 1-m.y. interval, using the depth boundaries established as described above.

In fitting the smooth curve to the data points, the lithologic core description from the *Initial Reports* was used as a guide. Single data points deviating more than 5 percent from adjacent ones were commonly disregarded except when the lithologic description indicated the likelihood of a significant change in $CaCO_3$ content. When a deflection was supported by two or more data points, the curve was fitted to them (Fig. 66 shows examples). The $CaCO_3$ curve for site 83 in Figure 66 shows large fluctuations over short distances in an otherwise uniform radiolarian nannofossil chalk. A more common case is represented by site 162 (Fig. 66), where large $CaCO_3$ variations are closely related to lithologic changes and single-point deviations appear to have no meaning.

RATES OF DEPOSITION

From the age-depth relationships, rates of deposition can be computed. In the simplest form, *sedimentation rate* in metres per million years for each 1-m.y. interval can be obtained by differentiations of the age-depth curve relative to age.

The sedimentation rate so defined has one obvious flaw: it does not take into account the compaction of sediment under load. Compaction with increasing overburden and, to a lesser extent, age reduces the porosity of the sedimentary deposits from initial values of 80 to 90 percent to final values of 50 to 30 percent or less. Thus, the sedimentation rate as defined above will tend to decrease with increasing age of the deposit. The decrease, moreover, does not show a simple correlation with age, because the porosity reduction is a complex function of the initial porosity, which depends strongly on the sediment type, thickness and density of the overburden, and diagenetic factors. Earlier investigations (for example, most *Initial Reports*) have either ignored the effect of compaction when determining sedimentation rates or have corrected for it by normalization to a standard porosity (for example, 50 percent; van Andel and Heath, 1973b). Although the values thus become less age dependent, the resulting sedimentation rates are wholly artificial, because sedimentary deposits with 50 percent porosity are not normally found on the deep-sea floor. A different number is thus needed to provide useful data: this is the *accumulation rate* in grams per square centimetre per thousand years.

Two sets of data in the *Initial Reports* can be used to compute average accumulation rates for each 1-m.y. interval: the wet bulk density obtained with the Gamma Ray Attenuation Porosity Evaluator (GRAPE), which includes a calculated porosity, and results of analyses of shore laboratory wet bulk density, porosity, and grain density. The techniques have been described by Evans and Cotterell (1970) and Bader (1970). A comprehensive discussion of the reliability of both methods is given by Bennett and Keller (1973). Carefully obtained GRAPE curves give good porosity values provided that reasonable grain-density numbers are introduced. Good laboratory analyses agree with the GRAPE data to ±5 percent. However, both procedures have often been carelessly executed; in particular, the laboratory data are of variable quality.

For the present study, we have used the GRAPE data, because their average quality is probably better than the other set, and they are far more continuous and available for many more cores. Apart from improper calibration, which can no longer be rectified, the principal sources of error are core distortion and the presence of unrecognized voids in the unopened core. These lead to bulk densities that are too low and porosities that are too high. Consequently, we fitted smooth curves to the data with emphasis on the highest reasonable wet bulk density and the lowest reasonable porosity values. These curves were fitted on plots obtained from the prime-data file at the Deep Sea Drilling Project, because the published graphs in the *Initial Reports* are on a very small scale and are sometimes erroneous. From these smooth curves, wet bulk density and porosity values were obtained at 2-m intervals beginning at 1 m below the sea floor. The upper metre was generally too distorted to be useful. Very high wet bulk density values due to the presence of nodules or chert were eliminated by comparison with the core description, as were values suspect because of voids or distortion.

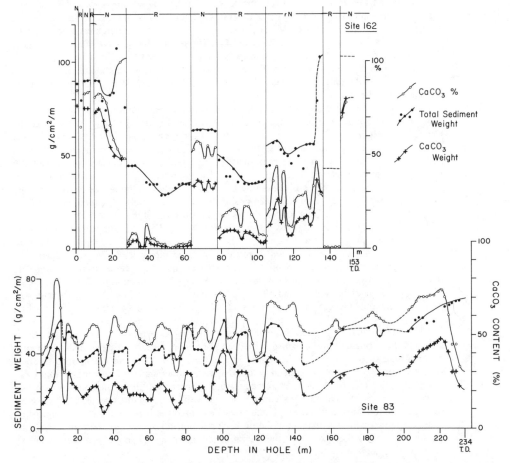

Figure 66. Examples of determination of accumulation rates in representative drill site (top) and drill site with high-frequency carbonate fluctuations (bottom). Open circles: carbonate content; black dots: total sediment weight (g/cm²/m); crosses: weight of calcareous fraction (g/cm²/m). Carbonate content scale in percent at right. Site 83 has a fairly uniform section of radiolarian nannofossil ooze changing downward to chalk. Lithology of site 162 shown at top (R, radiolarian ooze; N, nannofossil ooze or chalk; rN, radiolarian nannofossil ooze or chalk). Curves used for calculation of accumulation rates are fitted by hand as discussed in text.

From the digitized values the weight of solids per square centimetre per metre can be calculated as follows:

$$
\begin{aligned}
\text{wet weight of sediment} &= \text{wet weight of solids} + \text{weight of sea water}\\
\text{weight of sea water} &= \text{volume of sea water} \times \text{density of sea water}\\
\text{volume of sea water} &= \text{volume of voids} = \text{porosity}/100\\
\text{weight of sea water} &= 1.025 \times \text{porosity}/100 = 0.01025 \times \text{porosity}\\
\text{weight of solids} &= \text{wet weight of sediments} - \text{weight of sea water}\\
&= \text{wet bulk density} - 0.01025 \times \text{porosity}
\end{aligned}
$$

or, per metre of core,

$$100\,(\text{wet bulk density} - 0.01025 \times \text{porosity}).$$

Alternately, the weight of solids per metre could have been computed using the GRAPE wet bulk density values and grain densities determined in the laboratory. The latter, however, were judged too variable in quality and too sparse to be so used.

The weights of solids per metre of core were plotted against depth in the hole and a smooth curve fitted to the highest reasonable values. This smooth curve was digitized, and the data accumulated to provide accumulation rates in $g/cm^2/1,000$ yr for each 1-m.y. interval. Multiplication of the weights of solids per metre of core with the percentages of $CaCO_3$, fitting and digitizing a smooth curve on a plot of weight versus depth-in-hole, and accumulating the values for 1-m.y. intervals yield carbonate accumulation rates. Carbonate-free accumulation rates were then obtained from the difference between the bulk and carbonate accumulation rates. All accumulation rates as well as the sedimentation rates defined earlier have been tabulated in Appendix 3.

The accumulation rates are subject to errors of estimate resulting from the basic analytical data and from the processing procedure. The error of the GRAPE data is estimated by Bennett and Keller (1973) to be about ±5 percent for porosity and ±0.05 g/cm^3 for wet bulk density. Reading errors on the curves are less than 1 percent for porosity and $CaCO_3$ and less than 0.02 g/cm^3 for wet bulk density. The total error in the computation of the weight of solids per metre of core is then about ±10 $g/cm^2/m$.

Far more important is the error resulting from uncertainties in the age-depth relationship, because the first derivative of this relationship is the basis of all subsequent estimates of rates of deposition of various kinds. Errors in the position of the 1-m.y. boundaries can affect the intervening sediment thickness in a major way (Fig. 65). The effect can be considered in two different ways. In the first place, one can assume that the Berggren (1972a) time scale may be subject to nonrandom distortions of large dimensions. We have dealt with this assumption in Chapter 5. Secondly, one can postulate that the uncertainties of the absolute time scale are random and can be represented by an error function such as the one of Figure 64. For ages between 0 and 10 m.y., this function will define the uncertainties and thus also the possible errors of the rates of deposition. For ages greater than 10 m.y., the uncertainty at each 1-m.y. boundary is constrained by the uncertainty of the preceding and following boundaries. There is no satisfactory theory for errors in a time-series analysis in which the estimate of time is also subject to uncertainty. In the absence of such theory, it is reasonable to assume that the uncertainties at each boundary are constrained to some large fraction of the interval between two boundaries, for example, 67 percent. In that case, changes in the rate of deposition exceeding ±67 percent for the sedimentation rate and 75 percent for accumulation rates can be viewed with some confidence. The larger value for accumulation rates results from the additional error involved in this estimate. The consistent results obtained with calculations based on accumulation rates, such as those in the second section of Chapter 6, suggest that these confidence limits are far too conservative and that limits of ±10 percent are more realistic.

PALEODEPTH DETERMINATION

Menard (1969) and, more rigorously, Sclater and Francheteau (1970) have shown that the newly formed crust gradually and systematically subsides as it ages. This relationship between crustal age and basement depth was given a more quantitative form by Sclater and others

(1971) and has been successfully applied to paleotectonic problems by Sclater and McKenzie (1973), Anderson and Sclater (1972), and Anderson and Davis (1973). The same relationship can be used inversely to determine the subsidence history of a basement block for which the age and present depth are known (Berger, 1972, 1973a).

The paleodepth histories of the drill sites in the central equatorial Pacific have been computed by means of a procedure analogous to the one used by Berger (1973a). This procedure is based on the relationship between basement age and basement depth of the drill sites (van Andel and Bukry, 1973) rather than on regional correlations between sea-floor depth and geomagnetic polarity age (Sclater and others, 1971). This approach does not require a correction for the isostatic compensation resulting from loading by sediment, although the oldest paleodepths so obtained will be slightly too great, because the effect is not distributed over the entire subsidence history. This error is small, but corrections must be made for the difference between basement depth at the drill site and the mean basement depth in the vicinity, because many drill sites are located in places where the basement is anomalously deep.

Starting with the corrected basement depth obtained from van Andel and Bukry (1973), the paleodepth curve is obtained from a track parallel to the subsidence curve of Figure 67. The paleodepth values obtained for each 1-m.y. interval are then decreased by the thickness of sediment present at that time as given by the age–depth-in-hole relationship. Some drill sites did not reach basement. Here, the basement age, basement depth, and sediment thickness in the missing interval were estimated from the isochron and isopach maps of Figures 7 and 6B. The paleodepths for each 1-m.y. interval, obtained by calculation rather than graphically, are given in Appendix 3.

Confidence limits for paleodepth estimates are not easily estimated. The basic subsidence curve is only approximate, although the application of this concept in several paleotectonic studies has yielded consistent results. The curve is based on a generalized depth change; any particular point on the ocean floor may deviate by 50 to 200 m from the depth indicated for basement of that particular age. Depth variations of that amount over short distances of a few to a few tens of kilometres are common. The procedure followed here minimizes this variation by using a mean depth rather than the actual drill-site depth. Another error is introduced by the uncertainty limits of the estimate of basement age. Using the estimates of this error given by van Andel and Bukry (1973), this uncertainty appears to be on the

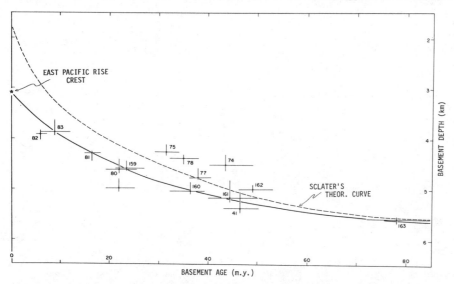

Figure 67. Relationship between basement age and basement depth in the central equatorial Pacific used for determination of paleodepth (after van Andel and Bukry, 1973). Horizontal and vertical bars at drill sites indicate errors of estimate. Theoretical curve after Sclater and others (1971).

order of ±50 to 100 m. Combined with the uncertainty resulting from local relief, this yields paleodepth confidence limits of approximately ±150 m.

PALEOLATITUDE AND PALEOLONGITUDE

The paleolatitudes and paleolongitudes listed in Appendix 3 have been calculated by rotating the drill sites back to their initial positions at the accreting plate edge in 1-m.y. steps. The rotation scheme used requires several tectonic assumptions discussed in Chapter 3 and has been derived from a fit of the axes of maximum equatorial deposition to the paleoequator. The procedure involves data from this study and is discussed in detail in Chapter 3.

SURFACE CORES

Data from a set of surface cores that penetrated pre-Quaternary deposits have been used for study of the temporal history of the dissolution of calcium carbonate (Chap. 4). The ages of these cores were determined or redetermined for this study on the basis of radiolarian assemblages; for a few, the ages came from Saito and others (1974). The basement age at each site was estimated from the basement isochron map of Figure 7, and the basement depth and thickness of sediment at each site were derived from Figure 6B. From these numbers the paleodepth, paleolatitude, and paleolongitude were computed according to the procedures outlined above. The data are presented in Appendix 5. They have larger uncertainties than those of Appendix 3 but can be used conservatively.

Appendix 2

Data From Sites 40 and 41

Sites 40 and 41 (McManus and Burns, 1970) are located close together near the northern margin of the area of study (App. 1, Fig. 63) in a region of brown clay and radiolarian ooze. Calcium carbonate content in cores from both sites is consistently below 1 percent (McManus and Burns, 1970, p. 438-439).

The two sites present serious biostratigraphic problems that prevent treatment of the data in a manner analogous to sites 42 to 163. The faunas are exclusively radiolarian with the exception of poor and altered nannofossil assemblages in cores 18 and 19 of site 40. The radiolarian biostratigraphy, revised by A. Sanfilippo and J. Westberg, is shown in Figure 68. Zone identifications for the nannofossil assemblages were made by D. Bukry. The data provide only a minimal absolute chronology with which sedimentation rates can be roughly estimated (Table 7). The calculation of accumulation rates is not justified.

The two sites present another problem. Notwithstanding their proximity, site 40 terminated in about 52-m.y.-old sedimentary deposits at an unknown distance above basement, whereas site 41 reached basalt overlain by 45- to 48-m.y.-old deposits (basement age of 48.5 m.y., according to van Andel and Bukry, 1973). Sedimentation rates at the two sites differ by a factor of 3 to 5. This large difference in basement age may be due to a large unconformity at the sediment-basement contact at site 41, to an intrusion mistakenly identified as basement at this same site, or to a large error in the assumed age of the lowermost sediments at

TABLE 7. SEDIMENTATION RATES AT SITES 40 AND 41

Age (m.y. B.P.)	Site 40		Site 41	
	Depth in hole (m)	Sedimentation rate (m/m.y.)	Depth in hole (m)	Sedimentation rate (m/m.y.)
40	?		?	
41	?	?	18	?
42	?	?	23	5.0
43	?	?	28.5	5.5
44	12	?	34	5.5
45	24	12	basement	
46	37	13	do.	
47	51	14	do.	
48	66	15	do.	
49	80	14	do.	
50	95	15	do.	
51	110	25	do.	
52	156	36	do.	

Note: Sedimentation rate listed against lower bound of interval.

site 40. The difference in sedimentation rates may be attributed to major downslope movement of sediments into the deep valley in which site 40 is located. These discrepancies call into question the use of the data from these two sites for regional synthesis purposes. Given the uncertainty of the basement age in the area, paleodepth histories have not been computed and paleolatitudes and paleolongitudes not tabulated, but a site-migration track based on an assumed age of 48 m.y. is shown in Figure 20.

At each site the middle Eocene section is overlain by 12 to 18 m of brown, nonfossiliferous clay of indeterminate age. This section may well contain major unconformities at the base or within the section.

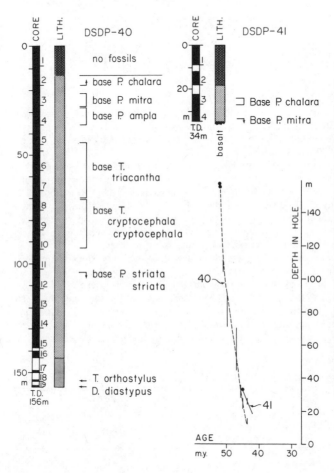

Figure 68. Graphic summaries for sites 40 and 41. Coring record on left with depth in hole in metres on left and core numbers on right. Black is recovered material. Symbols for lithologic column (LITH.) from legend to Appendix 3. Range of bases of radiolarian zones shown at right (from revisions by J. Westberg and A. Sanfilippo); calcareous nannofossil zone occurrences from D. Bukry. Lower right: age markers plotted against depth in hole. Vertical lines show age ranges of biostratigraphic markers; dashed and drawn lines represent smoothed age-versus-depth curves for sites. Sedimentation rates of Table 7 are derived from these curves.

Appendix 3

Tables and Graphs for Sites 42
through 163

Tables and graphs for sites 42 through 163 are on microfiche in pocket on back cover.

EXPLANATION OF TABLES

Procedures used to obtain the tabulated information have been described in detail in Appendix 1. Limited data from sites 40 and 41 are given in Appendix 2.

Sedimentation rates, accumulation rates, and calcium carbonate content are given as averages for each 1-m.y. interval; these are listed against the lower boundary of each interval. In the graphs, these values have been plotted against the midpoint of the appropriate interval.

Residual accumulation rates are the difference between the bulk and carbonate accumulation rates and represent the entire carbonate-free fraction.

Dotted patterns indicate unconformities and eroded sections. Blank space indicates lack of data or inadequate information.

Total depth (T.D.) is given when the drill did not end in basement.

EXPLANATION OF FIGURES

Time scale in millions of years is shown on left.

Core coverage, plotted against age, is shown in second column from left. Tick marks on left give depth in hole in tens of metres. Tick marks at right identify upper and lower core boundary. Black areas represent recovered core.

Third column shows lithology; symbols are explained on "Legend" page. Oblique boundary between units indicates gradual change; vertical boundary indicates alternating beds. Blank areas bordered by wavy lines represent unconformities.

Graphs at right show sedimentation and accumulation rates, calcium carbonate content, and paleodepth changes with age. Calibrations are at bottom of graph; curve identification in Legend.

Appendix 4

Biostratigraphic Tables for Sites 42 through 163

Biostratigraphic tables for sites 42 through 163 are on microfiche in pocket on back cover.

EXPLANATION OF TABLES

Biostratigraphic zone assignments are based on revisions of the biostratigraphic zonations presented in *Initial Reports* 5, 8, 9, and 16. For the sake of brevity, radiolarian and calcareous nannofossil zones are indicated with code numbers. Code numbers refer to the zonations of Moore (1971), Bukry (1973), and Bukry and others (1973); foraminiferal zonations are from Berggren (1972a). In the introductory tables, these zonations are correlated with the absolute chronology of Berggren (1972a), following the cited authors.

The drill-site tables show depth in the hole in metres on the left side of the column that indicates core recovery (black) and core numbers; zonations are to the right. A wavy line indicates a hiatus based on biostratigraphy, a drawn line indicates a firm zone boundary, and a dashed line signifies an interpolated zone boundary. Where cores are very widely spaced, no inferred zone boundaries are shown. Basement is indicated with a horizontal line across all three zonation columns.

Appendix 5

Tables for Surface Cores

Tables of data from surface cores are on microfiche in pocket on back cover.

EXPLANATION OF TABLES

Paleodepth values were estimated from basement ages (from Fig. 7) and basement depths obtained by adding present depth and sediment thickness (from Fig. 6B). Uncertainty of paleodepth estimate is ±150 m.

Basement age is estimated from Figure 7 to an accuracy of ±3 m.y. Sediment age is estimated from radiolarian assemblages (calcareous nannofossil assemblages for core JYNV 48P), using chronology of Moore (1971) and Bukry and others (1973). Lithology is from core descriptions of Scripps Institution of Oceanography and Lamont-Doherty Geological Observatory.

References Cited

Amos, A. F., Gordon, A. L., and Schneider, E. D., 1971, Water mass and circulation patterns in the region of the Blake-Bahamas Outer Ridge: Deep-Sea Research, v. 18, p. 145-165.

Anderson, R. N., 1974, Cenozoic motion of the Cocos plate relative to the asthenosphere and cold spots: Geol. Soc. America Bull., v. 85, p. 175-180.

Anderson, R. N., and Davis, E. E., 1973, A topographic interpretation of the Mathematician Ridge, Clipperton Ridge, East Pacific Rise system: Nature, v. 241, p. 191-193.

Anderson, R. N., and Sclater, J. G., 1972, Topography and evolution of the East Pacific Rise between 5° S and 20° S: Earth and Planetary Sci. Letters, v. 14, p. 433-441.

Arrhenius, G. O., 1952, Sediment cores from the East Pacific: Swedish Deep-Sea Exped. Rept. 1947-1948, v. 5, pt. 3, p. 189-201.

——1963, Pelagic sediments, in Hill, M. N., ed., The sea: New York, Interscience Pubs., Inc., v. 3, p. 655-727.

Atwater, Tanya, 1970, Implications of plate tectonics for the Cenozoic tectonic evolution of western North America: Geol. Soc. America Bull., v. 81, p. 3513-3536.

Bader, R. G., 1970, Shore-based laboratory procedures, in Bader, R. G., ed., Initial reports of the Deep Sea Drilling Project: Washington, D.C., U.S. Govt. Printing Office, v. 4, p. 745-753.

Baldwin, B., Coney, P. J., and Dickinson, W. R., 1974, Dilemma of a Cretaceous time scale and rates of sea-floor spreading: Geology, v. 2, p. 267-270.

Bandy, O. L., 1970, Upper Cretaceous-Cenozoic paleobathymetric cycles, eastern Panama and northern Colombia: Gulf Coast Assoc. Geol. Socs. Trans., v. 20, p. 181-193.

Bennett, R. H., and Keller, G. H., 1973, Physical properties evaluation, in van Andel, Tj. H., and Heath, G. R., eds., Initial reports of the Deep Sea Drilling Project: Washington, D.C., U.S. Govt. Printing Office, v. 16, p. 513-519.

Berger, W. H., 1968, Planktonic Foraminifera: Selective solution and paleoclimatic interpretation: Deep-Sea Research, v. 15, p. 31-43.

——1970a, Planktonic Foraminifera: Selective solution and the lysocline: Marine Geology, v. 8, p. 111-138.

——1970b, Biogenous deep-sea sediments; fractionation by deep-sea circulation: Geol. Soc. America Bull., v. 81, p. 1385-1402.

——1971, Sedimentation of planktonic Foraminifera: Marine Geology, v. 11, p. 325-358.

——1972, Deep-sea carbonates: Dissolution facies and age depth constancy: Nature, v. 236, p. 392-395.

——1973a, Cenozoic sedimentation in the eastern tropical Pacific: Geol. Soc. America Bull., v. 84, p. 1941-1954.

——1973b, Deep-sea carbonates: Evidence for a coccolith lysocline: Deep-Sea Research, v. 20, p. 917-921.

——1973c, Deep-sea carbonates: Pleistocene dissolution cycles: Foram. Research Jour., v. 3, p. 187-195.

Berger, W. H., and von Rad, Ulrich, 1972, Cretaceous and Cenozoic sediments from the Atlantic Ocean, in Hayes, D. E., and Pimm, A. C., eds., Initial reports of the Deep Sea Drilling Project: Washington, D.C., U.S. Govt. Printing Office, v. 14, p. 787-954.

Berger, W. H., and Winterer, E. L., 1974, Plate stratigraphy and the fluctuating carbonate line, *in* Hsü, K. J., and Jenkins, H. C., eds., Pelagic sediments on land and under the sea: Internat. Assoc. Sedimentologists Spec. Pub., v. 1, p. 11-48.

Berggren, W. A., 1969a, Rates of evolution in some Cenozoic planktonic Foraminifera: Micropaleontology, v. 15, p. 351-365.

——1969b, Cenozoic stratigraphic, planktonic foraminiferal zonation and the radiometric time scale: Nature, v. 224, p. 1072-1075.

——1971, Tertiary boundaries, *in* Funnell, B. F., and Riedel, W. R., eds., Micropaleontology of the oceans: New York, Cambridge Univ. Press, p. 693-809.

——1972a, A Cenozoic time scale: Some implications for geology and paleo-biogeography: Lethaia, v. 5, p. 195-215.

——1972b, Late Pliocene-Pleistocene glaciation, *in* Laughton, A. S., and Berggren, W. A., eds., Initial reports of the Deep Sea Drilling Project: Washington, D.C., U.S. Govt. Printing Office, v. 12, p. 953-965.

Berggren, W. A., and Van Couvering, J. A., 1974, The late Neogene biostratigraphy, geochronology and paleoclimatology of the last 15 million years in marine and continental sequences: Palaeogeography, Palaeoclimatology, Palaeoecology, v. 16, no. 1/2, 216 p.

Blakely, R. J., 1974, Geomagnetic reversals and crustal spreading rates during the Miocene: Jour. Geophys. Research, v. 79, p. 2979-2985.

Blow, W. H., 1969, Late middle Eocene to Recent planktonic foraminiferal biostratigraphy, *in* Bronniman, P., and Renz, H. H., eds., Proceedings of the International Conference on Planktonic Microfossils, Vol. 1: Leiden, E. J. Brill, 421 p.

Bode, G. W., and Cronan, D. S., 1973, Carbon and carbonate analyses, Leg 16, *in* van Andel, Tj. H., and Heath, G. R., eds., Initial reports of the Deep Sea Drilling Project: Washington, D.C., U.S. Govt. Printing Office, v. 16, p. 521-528.

Boyce, R. E., and Bode, G. W., 1973, Carbon and carbonate analyses, Leg 9, *in* Hays, J. D., ed., Initial reports of the Deep Sea Drilling Project: Washington, D.C., U.S. Govt. Printing Office, v. 9, p. 797-816.

Bramlette, M. N., 1961, Pelagic sediments, *in* Oceanography: Washington, D.C., Am. Assoc. Adv. Sci., p. 345-366.

——1965, Massive extinctions in biota at the end of Mesozoic time: Science, v. 148, p. 1696-1699.

Broecker, W. S., 1971, A kinetic model for the chemical composition of seawater: Quaternary Research, v. 1, p. 188-207.

Bukry, David, 1973, Coccolith stratigraphy, eastern equatorial Pacific, *in* van Andel, Tj. H., and Heath, G. R., eds., Initial reports of the Deep Sea Drilling Project: Washington, D.C., U.S. Govt. Printing Office, v. 16, p. 653-711.

Bukry, David, Dinkelman, M. G., and Kaneps, Ansis, 1973, Biostratigraphy of the equatorial East Pacific Rise, *in* van Andel, Tj. H., and Heath, G. R., eds., Initial reports of the Deep Sea Drilling Project: Washington, D.C., U.S. Govt. Printing Office, v. 16, p. 915-936.

Burckle, L. D., Ewing, J. I., Saito, Tsunemasa, and Leyden, R. R., 1967, Tertiary sediments from the East Pacific Rise: Science, v. 157, p. 537-540.

Burkov, V. A., 1969, The bottom circulation of the Pacific Ocean: Oceanology, v. 9, p. 179-188.

Chase, T. E., Menard, H. W., and Mammerickx, Jacqueline, 1970, Bathymetry of the North Pacific: Scripps Inst. Ocean. Inst. Marine Research Tech. Rept. TR-10.

Clague, D. A., and Dalrymple, G. B., 1973, Age of Kōko Seamount, Emperor seamount chain: Earth and Planetary Sci. Letters, v. 17, p. 411-415.

Clague, D. A., and Jarrard, R. D., 1973, Tertiary plate motion deduced from the Hawaiian-Emperor chain: Geol. Soc. America Bull., v. 84. p. 1135-1154.

Cook, H. E., 1972, Stratigraphy and sedimentation, *in* Hays, J. D., ed., Initial reports of the Deep Sea Drilling Project: Washington, D.C., U.S. Govt. Printing Office, v. 9, p. 933-943.

Cronan, D. S., van Andel, Tj. H., Heath, G. R., Dinkelman, M. G., Bennett, R. H., Bukry, D., Charleston, S., Kaneps, A., Rodolfo, K. S., and Yeats, R. S., 1972, Iron-rich sediments from the eastern equatorial Pacific, Leg 16, Deep Sea Drilling Project: Science, v. 175, p. 61-63.

Culberson, Charles, and Pytkowicz, R. M., 1968, Effect of pressure on carbonic acid, boric acid and the pH in seawater: Limnology and Oceanography, v. 13, p. 403–417.

Dalrymple, G. B., Lanphere, M. A., and Jackson, E. D., 1974, Contributions to the petrography and geochronology of volcanic rocks from the leeward Hawaiian Islands: Geol. Soc. America Bull., v. 85, p. 727–738.

Defant, A., 1961, Physical oceanography, Vol. 1: New York, Pergamon Press, Inc., 729 p.

Devereux, I., 1967, Oxygen isotope paleotemperature measurements on New Zealand Tertiary fossils: New Zealand Jour. Sci., v. 10, p. 988–1011.

Dewey, J. F., Pitman, W. C., III, Ryan, W.B.F., and Bonnin, Jean, 1973, Plate tectonics and the evolution of the alpine system: Geol. Soc. America Bull., v. 84, p. 3137–3180.

Dinkelman, M. G., 1973, Radiolarian stratigraphy, in van Andel, Tj. H., and Heath, G. R., eds., Initial reports of the Deep Sea Drilling Project: Washington, D.C., U.S. Govt. Printing Office, v. 16, p. 747–813.

Douglas, R. G., 1973, Evolution and bathymetric distribution of Tertiary deep-sea benthic Foraminifera: Geol. Soc. America Abs. with Programs, v. 5, p. 603.

Douglas, R. G., and Savin, S. M., 1971, Isotopic analyses of planktonic Foraminifera from the Cenozoic of the northwest Pacific, in Fischer, A. G., and Heezen, B. C., eds., Initial reports of the Deep Sea Drilling Project: Washington, D.C., U.S. Govt. Printing Office, v. 6, p. 1123–1127.

Dymond, J. R., 1966, Potassium-argon geochronology of deep-sea sediments: Science, v. 152, p. 1239–1241.

Dymond, J. R., Corliss, J. B., Heath, G. R., Field, C. W., Dasch, E. J., and Veeh, H. H., 1973, Origin of metalliferous sediments from the Pacific Ocean: Geol. Soc. America Bull., v. 84, p. 3355–3372.

Edmond, J. M., 1974, On the dissolution of carbonate and silicate in the deep ocean: Deep-Sea Research, v. 21, p. 455–480.

Evans, H. B., and Cotterell, C. H., 1970, Gamma ray attenuation density scanner, in Peterson, M.N.A., and Edgar, N. T., eds., Initial reports of the Deep Sea Drilling Project: Washington, D.C., U.S. Govt. Printing Office, v. 2, p. 460–471.

Ewing, J. I., Ewing, M., Aitken, T., and Ludwig, W. J., 1968, North Pacific sediment layers measured by seismic profiling: Am. Geophys. Union Geophys. Mon. 12, p. 147–173.

Francheteau, Jean, Harrison, C.G.A., Sclater, J. G., and Richards, M. L., 1970, Magnetization of Pacific seamounts: A preliminary polar curve for the northeastern Pacific: Jour. Geophys. Research, v. 75, p. 2035–2061.

Frerichs, W. E., 1970, Paleobathymetry, paleotemperature, and tectonism: Geol. Soc. America Bull., v. 81, p. 3445–3452.

Geitzenauer, K. R., Margolis, S. V., and Edwards, D. S., 1968, Evidence consistent with Eocene glaciation in a South Pacific deep-sea sedimentary core: Earth and Planetary Sci. Letters, v. 4, p. 173–177.

Gordon, A. L., and Gerard, R. D., 1970, North Pacific bottom potential temperature, in Hays, J. D., ed., Geological investigations of the North Pacific: Geol. Soc. America Mem. 126, p. 23–39.

Hawley, John, and Pytkowicz, R. M., 1969, Solubility of calcium carbonate in seawater at high pressure and 2° C: Geochim. et Cosmochim. Acta, v. 29, p. 1557–1561.

Hays, J. D., 1972, Initial reports of the Deep Sea Drilling Project: Washington, D.C., U.S. Govt. Printing Office, v. 9, 1205 p.

Hays, J. D., and Pitman, W. C., III, 1973, Lithospheric plate motion, sea level changes and climatic and ecological consequences: Nature, v. 246, p. 18–22.

Hays, J. D., Saito, T., Opdyke, N. D., and Burckle, L. H., 1969, Pliocene-Pleistocene sediments of the eastern equatorial Pacific; their paleomagnetic, biostratigraphic and climatic record: Geol. Soc. America Bull., v. 80, p. 1481–1514.

Hays, J. D., Cook, H., Jenkins, G., Orr, W., Goll, R., Cook, F., Milow, D., and Fuller, J., 1972, An interpretation of the geologic history of the eastern equatorial Pacific from the drilling results of the Glomar Challenger, Leg 9, in Hays, J. D., ed., Initial reports of the Deep Sea Drilling Project: Washington, D.C., U.S. Govt. Printing Office, v. 9, p. 909–931.

Heath, G. R., 1969a, Carbonate sedimentation in the abyssal equatorial Pacific during the past 50 million years: Geol. Soc. America Bull., v. 80, p. 689–694.

Heath, G. R., 1969b, Mineralogy of Cenozoic deep-sea sediments from the equatorial Pacific Ocean: Geol. Soc. America Bull., v. 80, p. 1997–2018.

——1974, Dissolved silica and deep-sea sediments, in Hay, W. W., ed., Studies in paleo-oceanography: Soc. Econ. Paleontologists and Mineralogists Spec. Pub. 20, p. 77–93.

Heath, G. R., and Culberson, Charles, 1970, Calcite: Degree of saturation, rate of dissolution and the compensation depth in the deep ocean: Geol. Soc. America Bull., v. 81, p. 3157–3160.

Heath, G. R., and van Andel, Tj. H., 1973, Tectonics and sedimentation in the Panama Basin; geologic results of Leg 16, Deep Sea Drilling Project, in van Andel, Tj. H., and Heath, G. R., eds., Initial reports of the Deep Sea Drilling Project: Washington, D.C., U.S. Govt. Printing Office, v. 16, p. 899–913.

Heath, G. R., Moore, T. C., Jr., Dauphin, J. P., and Opdyke, N. D., 1975, Quartz, organic carbon, opal, and calcium carbonate in Holocene, 600,000-year, and Brunhes-Matuyama age pelagic sediments of the North Pacific: Geol. Soc. America Bull. (in press).

Heirtzler, J. R., Dickson, G. O., Herron, E. M., Pitman W. C., III, and Le Pichon, X., 1968, Marine magnetic anomalies, geomagnetic field reversals and motions of the ocean floor and continents: Jour. Geophys. Research, v. 73, p. 2119–2136.

Herron, E. M., 1972, Sea-floor spreading and the Cenozoic history of the east-central Pacific: Geol. Soc. America Bull., v. 83, p. 1671–1692.

Hey, R. N., Deffeyes, K. S., Johnson, G. L., and Lowrie, A., 1972, The Galapagos triple junction and plate motions in the eastern Pacific: Nature, v. 237, p. 20–22.

Hollister, C. D., Johnson, D. A., and Lonsdale, P. F., 1974, Current-controlled abyssal sedimentation: Samoa Passage, equatorial West Pacific: Jour. Geology, v. 82, p. 275–300.

Hornibrook, N. de B., 1968, Distribution of some warm-water Foraminifera in the New Zealand Tertiary: Tuatara, v. 16, p. 11–15.

Jackson, E. D., Silver, E. A., and Dalrymple, G. B., 1972, Hawaiian-Emperor chain and its relation to Cenozoic circum-Pacific tectonics: Geol. Soc. America Bull., v. 83, p. 601–618.

Johnson, D. A., 1972a, Eastward flowing bottom currents along the Clipperton Fracture Zone: Deep-Sea Research, v. 19, p. 253–257.

——1972b, Ocean-floor erosion in the equatorial Pacific: Geol. Soc. America Bull., v. 83, p. 3121–3144.

Johnson, D. A., and Johnson, T. C., 1970, Sediment redistribution by bottom currents in the central Pacific: Deep-Sea Research, v. 17, p. 157–170.

Jones, E.J.W., Ewing, M., Ewing, J. I., and Eittreim, S. L., 1970, Influences of Norwegian Sea overflow water on sedimentation in the northern North Atlantic and Labrador Sea: Jour. Geophys. Research, v. 75, p. 1655–1680.

Kaneps, Ansis, 1970, Late Neogene biostratigraphy (planktonic Foraminifera) biogeography and depositional history [Ph.D dissert.]: New York, Columbia Univ., 185 p.

Kennett, J. P., and Brunner, C. A., 1973, Antarctic late Cenozoic glaciation; evidence for initiation of ice-rafting and inferred increased bottom water activity: Geol. Soc. America Bull., v. 84, p. 2043–2052.

Kennett, J. P., Burns, R. E., Andrews, J. E., Churkin, M., Jr., Davies, T. A., Dumitrica, P., Edwards, A. R., Galehouse, J. S., Packham, G. H., and van der Lingen, G. J., 1972, Australian-Antarctic continental drift, paleo-circulation changes and Oligocene deep-sea erosion: Nature Phys. Sci., v. 239, p. 51–55.

Kennett, J. P., Houtz, R. E., Andrews, P. V., Edwards, A. R., Gostin, V. A., Hajos, N., Hampton, M., Jenkins, D. G., Margolis, S. V., Ovenshine, A. T., and Perch-Nielsen, K., 1975, Cenozoic paleo-oceanography in the southwest Pacific Ocean and the development of the circum-Antarctic current, in Kennett, J. P., and Houtz, R. E., eds., Initial reports of the Deep Sea Drilling Project: Washington, D.C., U.S. Govt. Printing Office, v. 29, p. 1155–1169.

Knauss, J. A., 1962, On some aspects of the deep circulation of the Pacific: Jour. Geophys. Research, v. 67, p. 3943–3954.

Koblentz-Mishke, O. J., Volkovinsky, V. V., and Kabanova, J. G., 1970, Plankton primary production of the world ocean, in Wooster, W. S., ed., Scientific exploration of the

South Pacific: Washington, D.C., Natl. Acad. Sci., p. 183-193.

Larson, R. L., and Chase, C. G., 1970, Relative velocities of the Pacific, North American and Cocos plates in the Middle America region: Earth and Planetary Sci. Letters, v. 7, p. 425-432.

Larson, R. L., and Pitman, W. C., III, 1972, World-wide correlation of Mesozoic magnetic anomalies and its implications: Geol. Soc. America Bull., v. 83, p. 3645-3662.

Larson, R. L., Moberly, R., Bukry, D., Foreman, H., Garner, J. V., Keene, J., Lancelot, Y., Luterbacher, H., Marshall, M., and Matter, A., 1973, Leg 32, Deep Sea Drilling Project: Geotimes, v. 18, December, p. 14-17.

Le Pichon, X., Ewing, M., and Truchan, M., 1971, Sediment transport and distribution in the Argentine Basin. 2, Antarctic bottom current passage into the Brazil Basin: Phys. Chem. Earth, v. 8, p. 49-77.

Li, Y. H., Takahashi, T., and Broecker, W. S., 1969, Degree of saturation of $CaCO_3$ in the oceans: Jour. Geophys. Research, v. 74, p. 5507-5525.

Lisitsin, A. P., 1970, Sedimentation and chemical considerations, in Wooster, W. S., ed., Scientific exploration of the South Pacific: Washington, D.C., Natl. Acad. Sci., p. 89-132.

——1972, Sedimentation in the world ocean: Soc. Econ. Paleontologists and Mineralogists Spec. Pub. 17, 218 p.

Luyendyk, B. P., Forsyth, D., and Phillips, J. D., 1972, Experimental approach to the paleocirculation of the oceanic surface waters: Geol. Soc. America Bull., v. 83, p. 2649-2664.

Malfait, B. T., and Dinkelman, M. G., 1972, Circum-Caribbean tectonic and igneous activity and the evolution of the Caribbean plate: Geol. Soc. America Bull., v. 83, p. 251-272.

Mammerickx, Jacqueline, Anderson, R. N., Menard, H. W., and Smith, S. M., 1975, Morphology and tectonic evolution of the east-central Pacific: Geol. Soc. America Bull., v. 86, p. 111-118.

Margolis, S. V., and Kennett, J. P., 1970, Antarctic glaciation during the Tertiary recorded in sub-Antarctic deep-sea cores: Science, v. 170, p. 1085-1087.

——1971, Cenozoic paleoglacial history of Antarctica recorded in sub-Antarctic deep-sea cores: Am. Jour. Sci., v. 271, p. 1-36.

Maxwell, A. E., Von Herzen, R. P., Hsü, K. L., Andrews, J. E., Saito, T., Percival, S. E., Milow, E. D., and Boyce, R. E., 1970, Deep sea drilling in the South Atlantic: Science, v. 168, p. 1047-1059.

McManus, D. A., and Burns, R. E., eds., 1970, Initial reports of the Deep Sea Drilling Project: Washington, D. C., U.S. Govt. Printing Office, v. 5, 827 p.

Menard, H. W., 1969, Elevation and subsidence of the oceanic crust: Earth and Planetary Sci. Letters, v. 6, p. 275-284.

——1973, Depth anomalies and the bobbing motion of drifting islands: Jour. Geophys. Research, v. 78, p. 5128-5138.

Minster, J. B., Jordan, T. H., Molnar, Peter, and Haines, E., 1974, Numerical modeling of instantaneous plate tectonics: Royal Astron. Soc. Geophys. Jour., v. 36, p. 541-576.

Moberly, R., 1972, Origin of lithosphere behind island arcs with reference to the western Pacific, in Shagam, R., ed., Studies in earth and space sciences: Geol. Soc. America Mem. 132, p. 35-56.

Molnar, Peter, and Atwater, Tanya, 1973, Relative motion of hot spots in the mantle: Nature, v. 246, p. 288-291.

Moore, T. C., Jr., 1970, Abyssal hills in the central equatorial Pacific: Sedimentation and stratigraphy: Deep-Sea Research, v. 17, p. 573-593.

——1971, Radiolaria, in Tracey, J. I., and Sutton, G. B., eds., Initial reports of the Deep Sea Drilling Project: Washington, D.C., U.S. Govt. Printing Office, v. 8, p. 727-775.

——1972, DSDP: Successes, failures, proposals: Geotimes, v. 17, no. 7, p. 27-31.

Moore, T. C., Jr., Heath, G. R., and Kowsmann, R. O., 1973, Biogenic sediments of the Panama Basin: Jour. Geology, v. 81, p. 458-472.

Moore, T. C., Jr., van Andel, Tj. H., Sancetta, C., and Pisias, N., 1975, Cenozoic hiatuses in pelagic sediments, in Riedel, W. R., and Saito, T., Marine plankton and sediments: New York, Am. Mus. Nat. History Micropaleontology Press (in press).

Morgan, W. J., 1972, Plate motions and deep mantle convection, in Shagam, R., ed., Studies in earth and space sciences: Geol. Soc. America Mem. 132, p. 7–32.

Murray, J., and Renard, A. F., 1891, Deep-sea deposits: London, Challenger Exped. Rept., 525 p.

Olausson, Eric, 1960, Descriptions of sediment cores from the central and western Pacific with adjacent Indonesian region: Swedish Deep-Sea Exped. Rept. 1947–1948, v. 6, pt. 5, p. 161–168.

Page, R. W., and McDougall, Ian, 1970, Potassium-argon dating of the Tertiary Fl-2 stage in New Guinea and its bearing on the geological time scale: Am. Jour. Sci., v. 269, p. 321–342.

Panfilova, S. G., 1967, Bottom temperature and salinity in the Pacific Ocean: Oceanology, v. 7, p. 690–694.

Parker, F. L., and Berger, W. H., 1971, Faunal and solution patterns of planktonic Foraminifera in surface sediments of the South Pacific: Deep-Sea Research, v. 18, p. 73–107.

Peterson, M.N.A., 1966, Calcite rates of dissolution in a vertical profile in the central Pacific: Science, v. 154, p. 1542–1544.

Phillips, J. D., and Forsyth, D., 1972, Plate tectonics, paleomagnetism and the opening of the Atlantic: Geol. Soc. America Bull., v. 83, p. 1579–1600.

Pimm, A. G., 1974, Sedimentology and history of the northeastern Indian Ocean from late Cretaceous to Recent, in von der Borch, C. C., and Sclater, J. G., eds., Initial reports of the Deep Sea Drilling Project: Washington, D.C., U.S. Govt. Printing Office, v. 22, p. 771–773.

Pisias, N. G., 1974, Model of late Pleistocene–Holocene variations in rate of sediment accumulation: Panama Basin, eastern equatorial Pacific [M.S. thesis]: Corvallis, Oregon State Univ., 77 p.

Pitman, W. C., III, and Talwani, Manik, 1972, Sea-floor spreading in the North Atlantic: Geol. Soc. America Bull., v. 83, p. 619–646.

Rea, D. K., 1974, Tectonics of the East Pacific Rise, 5° to 12° South latitude [Ph.D. dissert.]: Corvallis, Oregon State Univ., 139 p.

Rea, D. K., and Malfait, B. T., 1974, Evolution of the northern Nazca plate: Geology, v. 2, p. 317–320.

Reid, J. L., 1962, On the circulation, phosphate-phosphorus content and zooplankton volumes in the upper part of the Pacific Ocean: Limnology and Oceanography, v. 7, p. 287–306.

Revelle, R. R., 1944, Marine bottom samples collected in the Pacific by the Carnegie on its seventh cruise: Carnegie Inst. Washington Pub. 556, 182 p.

Riedel, W. R., and Funnell, B. M., 1964, Tertiary sediment cores and microfossils from the Pacific Ocean floor: Geol. Soc. London Quart. Jour., v. 120, p. 305–368.

Rodda, P., Snelling, N. J., and Rex, D. C., 1967, Radiometric age data on rocks from Viti Levu, Fiji: New Zealand Jour. Geology and Geophysics, v. 10, p. 1248–1259.

Rona, P. A., 1973, Worldwide unconformities in marine sediments related to eustatic changes of sea level: Nature Phys. Sci., v. 244, p. 25–26.

Ruddiman, W. F., 1972, Sediment redistribution on the Reykjanes Ridge; seismic evidence: Geol. Soc. America Bull., v. 83, p. 2039–2062.

Ryther, J. H., 1963, Geographic variations in productivity, in Hill, M. N., ed., The sea, Vol. 2: New York, Interscience Pubs., Inc., p. 347–380.

Saito, T., Burckle, L. D., and Hays, J. D., 1974, Implications of some pre-Quaternary cores and dredgings, in Hay, W. W., Studies in paleo-oceanography: Soc. Econ. Paleontologists and Mineralogists Spec. Pub. 20, p. 6–36.

Schlanger, S. O., Jackson, E. D., Boyce, R. E., Cook, H. E., Jenkyns, H., Johnson, D. A., Kaneps, A., Kelts, K., Martini, E., McNulty, C. L., and Winterer, E. L., 1974, Leg 33, Deep Sea Drilling Project, testing a hot spot theory: Geotimes, v. 19, no. 3, p. 16–20.

Schneider, E. D., and Heezen, B. C., 1966, Sediments of the Caicos outer ridge: Geol. Soc. America Bull., v. 77, p. 1381–1398.

Sclater, J. G., and Francheteau, Jean, 1970, The implications of terrestrial heat flow observations for current tectonic and geochemical models of the crust and upper mantle of the earth: Royal Astron. Soc. Geophys. Jour., v. 20, p. 509–542.

Sclater, J. G., and McKenzie, D. P., 1973, Paleobathymetry of the South Atlantic: Geol. Soc. America Bull., v. 84, p. 3203-3216.

Sclater, J. G., Anderson, R. N., and Bell, M. L., 1971, Elevation of ridges and evolution of the central eastern Pacific: Jour. Geophys. Research, v. 76, p. 7888-7915.

Shackleton, N. J., and Kennett, J. P., 1975a, Paleotemperature history of the Cenozoic and the initiation of antarctic glaciations; oxygen and carbon isotope analyses at D.S.D.P. sites 277, 279, and 281, *in* Kennett, J. P., and Houtz, R. E., eds., Initial reports of the Deep Sea Drilling Project: Washington, D.C., U.S. Govt. Printing Office, v. 29, p. 743-755.

——1975b, Late Cenozoic oxygen and carbon isotopic changes at D.S.D.P. site 284; implications for glacial history of the northern hemisphere and Antarctica, *in* Kennett, J. P., and Houtz, R. E., eds., Initial reports of the Deep Sea Drilling Project: Washington, D.C., U.S. Govt. Printing Office, v. 29, p. 801-808.

Shaw, H. R., 1973, Mantle convection and volcanic periodicity in the Pacific; evidence from Hawaii: Geol. Soc. America Bull., v. 84, p. 1505-1526.

Shaw, H. R., and Jackson, E. D., 1973, Linear island chains in the Pacific; results of thermal plumes or gravitational anchors?: Jour. Geophys. Research, v. 78, p. 8634-8652.

Stommel, H., and Arons, A. B., 1960, On the abyssal circulation patterns in oceanic basins: Deep-Sea Research, v. 6, p. 217-233.

Tracey, J. I., and Sutton, G. D., eds., 1971, Initial reports of the Deep Sea Drilling Project: Washington, D.C., U.S. Govt. Printing Office, v. 8, 1037 p.

Tracey, J. I., Sutton, G. D., Nesteroff, W. D., Galehouse, J. S., von der Borch, C. C., Moore, T. C., Jr., Bilal-ul-haq, U. Z., and Beckmann, L. P., 1971, Leg 8 summary, *in* Tracey, J. I., and Sutton, G. D., eds., Initial reports of the Deep Sea Drilling Project: Washington, D.C., U.S. Govt. Printing Office, v. 8, p. 17-42.

Truchan, Marek, and Larson, R. L., 1972, Tectonic lineaments on the Cocos plate: Earth and Planetary Sci. Letters, v. 17, p. 426-432.

Turner, D. L., 1970, Potassium-argon dating of Pacific coast Miocene foraminiferal stages: Geol. Soc. America Spec. Paper 124, p. 91-126.

van Andel, Tj. H., 1972, Establishing the age of the oceanic basement: Comments on Earth Sci.: Geophysics, v. 2, p. 157-168.

——1973, Texture and dispersal of sediments in the Panama Basin: Jour. Geology, v. 81, p. 434-457.

van Andel, Tj. H., and Bukry, David, 1973, Basement ages and basement depths in the eastern equatorial Pacific: Geol. Soc. America Bull., v. 84, p. 2361-2370.

van Andel, Tj. H., and Heath, G. R., eds., 1973a, Initial reports of the Deep Sea Drilling Project: Washington, D.C., U.S. Govt. Printing Office, v. 16, 949 p.

——1973b, Geological results of Leg 16: The central equatorial Pacific west of the East Pacific Rise, *in* van Andel, Tj. H., and Heath, G. R., eds., Initial reports of the Deep Sea Drilling Project: Washington, D.C., U.S. Govt. Printing Office, v. 16, p. 937-949.

van Andel, Tj. H., and Moore, T. C., Jr., 1974, Cenozoic calcium carbonate distribution and calcite compensation depth in the central equatorial Pacific: Geology, v. 2, p. 87-92.

van Andel, Tj. H., Heath, G. R., Malfait, B. T., Heinrichs, D. E., and Ewing, J. I., 1971, Tectonics of the Panama Basin, eastern equatorial Pacific: Geol. Soc. America Bull., v. 82, p. 1489-1508.

Veevers, J. J., 1969, Paleogeography of the Timor Sea region: Palaeogeography, Palaeoclimatology, Palaeoecology, v. 6, p. 125-140.

Vogt, P. E., Avery, O. E., Schneider, E. D., Anderson, C. N., and Bracey, D. R., 1969, Discontinuities and seafloor spreading: Tectonophysics, v. 8, p. 285-318.

Weissel, J. K., and Hayes, D. E., 1972, Magnetic anomalies in the southeast Indian Ocean, *in* Hayes, D. E., ed., Antarctic oceanology II: The Australian-New Zealand sector: Am. Geophys. Union Antarctic Research Ser., v. 19, p. 165-196.

Weyl, Richard, 1973, Die palaeogeographische Entwicklung Mittelamerikas: Zentralbl. Geologie und Paläontologie, v. 1, p. 433-466.

Winterer, E. L., 1973, Sedimentary facies and plate tectonics of the equatorial Pacific: Am. Assoc. Petroleum Geologists Bull., v. 57, p. 265-282.

Winterer, E. L., and Ewing, J. I., eds., 1973, Initial reports of the Deep Sea Drilling Project: Washington, D.C., U.S. Govt. Printing Office, v. 17, 930 p.

Winterer, E. L., and Riedel, W. R., eds., 1971, Initial reports of the Deep Sea Drilling Project: Washington, D.C., U.S. Govt. Printing Office, v. 7, pt. 1, 1841 p.; pt. 2, 916 p.

Wüst, G., 1929, Schichtung und Tiefenzirkulation des Pazifischen Ozeans: Veröffentlichungen Inst. für Meereskunde Univ. Berlin, N.F. (A. Geographie und Naturwiss.), v. 20, p. 1-64.

Manuscript Received by the Society September 12, 1974

Revised Manuscript Received November 18, 1974

Manuscript Accepted December 5, 1974

Present Address: (Heath and Moore) Graduate School of Oceanography, University of Rhode Island, Kingston, Rhode Island 02881